Pole Dancing, Empowerment and Embodiment

Also by Samantha Holland

ALTERNATIVE FEMININITIES: Body, Age and Identity

REMOTE RELATIONSHIPS IN A SMALL WORLD (*edited*)

Pole Dancing, Empowerment and Embodiment

Samantha Holland
Leeds Metropolitan University, UK

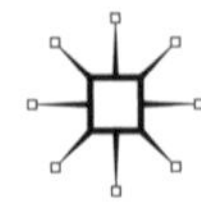

First published 2010 by
PALGRAVE MACMILLAN

Palgrave Macmillan in the UK is an imprint of Macmillan Publishers Limited, registered in England, company number 785998, of Houndmills, Basingstoke, Hampshire RG21 6XS.

Palgrave Macmillan in the US is a division of St Martin's Press LLC, 175 Fifth Avenue, New York, NY 10010.

Palgrave Macmillan is the global academic imprint of the above companies and has companies and representatives throughout the world.

ISBN 978–0–230–21038–7 hardback

This book is printed on paper suitable for recycling and made from fully managed and sustained forest sources. Logging, pulping and manufacturing processes are expected to conform to the environmental regulations of the country of origin.

A catalogue record for this book is available from the British Library.

A catalog record for this book is available from the Library of Congress.

10 9 8 7 6 5 4 3 2 1
19 18 17 16 15 14 13 12 11 10

Printed and bound in Great Britain by
CPI Antony Rowe, Chippenham and Eastbourne

Contents

List of Figures

Acknowledgements

I would like to thank the 37 polers who I interviewed in the UK, Sydney and New York. Every one of them was very kind, friendly, courteous and reflexive about their experiences, and I appreciate the time they gave up to talk to me and their willingness to share their stories. Their names/pseudonyms are listed in Chapter 1 and website links are included in Appendix 2, but I would like to particularly thank Jennifer and Bobbi, who not only gave up their time to talk to me but also put me in touch with their students; to Alison, who gave me a Pole People DVD; to Genevieve who collected me from the station and drove me to her studio; to KT and to Elena who both schlepped into the centre of London to meet me; to Anne and Shelly who were, and are, welcoming and helpful; to Maria and Eva who gave up their lunch hours; to James who gave up his Saturday morning and invited me to watch the Aussie Pole Boys perform at Slide in Sydney; and to Natalie Te Kanawa, Mia Gilson and Louisa at Pole Candy who contacted me about the online pole questionnaire. I am also grateful to the 132 women and 3 men who completed my questionnaire, and to Andy who answered further questions by email.

Photographs in this book, including the cover, were very kindly provided by Jennifer Critelli, Shelly Glass, Alison Hudd, Anne Noakes and Jason Parlour.

Thank you to:

- Feona Attwood, Sheila Scraton and Julie Harpin for their friendship and support, and to Feona for reading and commenting on the chapters;
- the many people, friends and friends of friends, who sent me cuttings or emails with references to pole classes and pole dancing; or who found themselves in the middle of a conversation about pole classes. They include Joan Phillip, Rod Allen, Jo Hammond, John Payne, Renyi Lim, Yrla Virvelpike, Emma Mitchell, Cathy Killick, Teela Sanders, Anne Flintoff, Liz Green, Joe Cook, Kristyn Gorton and Elaine Campbell;
- Sofia Aboim in Lisbon with whom I shared the joys, but mainly the pains, of writing;

- the very patient and always pleasant Olivia Middleton and the anonymous reviewer at Palgrave Macmillan, both of whom were encouraging and helpful.
- Leeds Metropolitan University which funded the data collection for this study.

Love and thanks to my family: my mum and stepdad, Jo and Mick Hammond, my uncle Charles 'Harbie' Holland, and my dad and stepmother, Kevin and Mary Callaghan.

And last, but not least, all my love to Sam Hinchliffe, for he is awesome.

Introduction

> *For feminist scholars, the body has always been – and continues to be – of central importance for understanding women's embodied experiences and practices and cultural and historical constructions of the female body in the various contexts of social life.*
>
> (Davis, 1997, p. 7)

> JESS: *A lot of them have bought poles.*
> SH: *It is amazing how many people find room for poles.*
> JESS: *It is, yes: 'I need a pole – let's move the bed!'*

Imagine a female athlete on a horizontal bar: she wears appropriate clothing (usually only a leotard); her display is acrobatic and technical, borne of hours of practice and requires disciplined physical strength, flexibility and skill. She demonstrates upper body and core strength, balance and physical confidence; she performs moves which require her to do the splits only in mid-air; and she has to be able to perform her moves upside down. All this is performed to music. Now change the word 'horizontal' for 'vertical' and you have, in essence, a poler.

This book aims to provide an up to date, international, multidisciplinary empirical account of a particular sort of gendered leisure activity and how it fits into wider discourses about bodies, gender, age and fitness. This activity is the relatively recent phenomenon of pole classes, also called, for instance, pole fitness, pole dancing, polenastic, vertical fitness, cardio pole, pole-da-cise, vertical dance, pole exercise and polebatics (new to Germany, Austria and Switzerland),[1] which are now held in many towns and cities across the UK, and in many other countries

Canada, the Netherlands, France, Australia and New 'ountries worldwide). In particular, I want to explore 'e agency and espouse liberation and, sometimes, .ɛrment through something as seemingly problematic .ɑsses. Pole classes are of particular interest because they fit ..ʋ many current academic and cultural debates about shifts in society (particularly around sexualities) and changes in leisure and exercise. For example, pole addresses issues about the sexualisation of culture, body image, gendered leisure time and exercise fashions. This book is the first of its kind in examining the worldwide phenomenon of pole classes, drawing on data collected via a variety of methods and locations – these are discussed in full in Chapter 1. Briefly though, I was a participant and/or observer at pole classes and conducted a total of 37 interviews with instructors and students of pole dancing in various places in the UK and in Sydney and New York. The participants were aged between 18 and 60 years, were of a wide variety of body sizes and although primarily white, included South Asian and African American participants. I also sent an online questionnaire to 9 pole schools worldwide for their members to complete, which was forwarded to message boards or individuals, and which resulted in 135 responses worldwide. The questionnaire data provides an international overview of who is doing pole and why and where, and what pole means to the women who do it, alongside the more detailed stories gleaned during the qualitative data collection.

Before I began this research I had, I now realise, a tenuous grasp of what pole classes, and pole dancing, really were. And, with the greatest respect, I imagine this is the case with many people. My understanding came from media or film images, such as the pole dancers in the background of TV shows like The Sopranos or The Wire, near-naked, bored and pouting, and not really doing much more than walking around or leaning on the pole (and, don't forget, probably those actresses were not polers) or unclear stills of Demi Moore or Kate Moss from film or music videos. One participant, Genevieve, called women who just walk around the pole 'moochers' because they do not do tricks; to mooch is to wander about aimlessly, which is not an accurate portrayal of a pole routine. I knew that pole dancing happened in lap-dancing clubs, I knew I hadn't seen any men doing it (this was in 2005) and I knew women didn't seem to wear much when they did it. All these things, actually, are, to some extent, true. But these are also not true in that pole classes are an offshoot of work within, or on the periphery of, the sex industry and so are different. They are not subject to the male gaze because pole classes are, for the vast majority, all-female; they may be about sexiness to some extent but they are not necessarily about sex, and most exercise classes per se are about looking or feeling 'sexy' so we could also infer that they are about feeling more confident about one's body; and the classes are about becoming physically strong and most pole instructors and schools pride themselves on their positive, affirmative attitude to body image, fun and achievement. So to properly imagine a pole class you must summon up a picture of a group of women, often red-faced and sweating, of varying ages and sizes, the sound of music, swearing, laughter, talking and occasional grunts of exertion. It is mostly not glamorous. But pole has caught the public imagination, and its image is not always accurate, and not always positive. It prompts a passionate response in those who criticise it as being part of the sex industry and therefore are demeaning to those who do it (this is a mix of journalists and academics but more distinction between them will be drawn later) and an equally impassioned reaction from those who actually do it.

A 2005/6 UK television advertisement for a men's body spray showed a young, skinny man spraying a coat stand which a woman then obligingly slid up and down, proving how irresistible the spray would make the man. In 2005 the Art Gallery of Ontario exhibited Henry Moore sculptures and had large line drawings of pole dancers on the walls (artist unknown). In late 2008 the Behind the Shutters gallery in London held an exhibition which included pole-dancing robots. The robots, made by Giles Walker, were made from materials found in scrap yards and operated via a PC. In early 2009 the final episode of Battlestar Galactica had several scenes set in a pole-dancing club with a split-second shot of a real poler doing a 'Flag' move and in spring 2009 a UK poler called Deb Riley accompanied DJ Talent on the semi-finals of the television programme, *Britain's Got Talent*.[2] Also in 2009 there was an international media outcry because Sam Remmer, another UK poler, had done a gymnastic, balletic display in a Plymouth high school – the display was performed using a pole (I mention this again in Chapter 6); and again in 2009 a pole school in Essex had its windows smashed simply because it offered pole classes. There is something about pole which piques our interest, as Catherine M. Roach recounts:

> It's the pole work that more and more comes to fascinate me. 'Pole work', I learn, is the industry term for a dancer's use in stage routines of the upright metal poles – chrome, brass, stainless steel or enamel, usually two inches in diameter, and bolted to the floor and ceiling. Pole work is a speciality. Not all dancers do it. While I never visit a

> club that doesn't have a pole, not all clubs feature true pole work. All dancers will at the very least put a hand up to the pole as they move from one side of the stage to another, using it to steady themselves or to lean out and swing their hair. ... Only some, however, are true mistresses of the pole. Developing a full repertoire of pole tricks requires months, if not years, of practice. ... Wow, I think, now this – theoretically – I would like to learn. Marie tells me that it takes a great deal of upper body strength, of which I sadly have none, and a lot of coordination and practice as well.
>
> (Roach, 2007, pp. 30–1)

The following chapters will examine why women began pole classes and their experiences of them. It quickly became apparent to me that pole classes have a lot to tell us about gender, embodiment, the mainstreaming of sex, women's leisure time, women's exercise, agency, feminism, ageing and even about modern masculinity. The main threads which run throughout the book are gendered embodiment and body image and its manifestations through cultural fashions such as exercise classes. I will also explore the paradoxical nature of pole classes; and interrogate the feelings of physical empowerment and liberation experienced by pole students and the possibilities or limitations of these feelings.

The recurring theme for some of the participants, primarily students of pole rather than the instructors, that emerged unexpectedly was a history of disliking physical exercise. In part it appeared to attract women because it is called 'dancing' rather than exercise and because of its safely feminised, even sexualised, image. Conversely, the participants' accounts of feeling liberated and even empowered by pole exercise are also key to understanding its popularity; many of them felt that pole classes offered a space to resist gendered, embodied expectations of them, and it is these two apparently opposing (some might say confounding) findings which inform the following chapters.

Why did I choose this subject? This research feels in many ways to follow on from previous work I have done about women, ageing and alternative femininities (Holland, 2004) in that, again, I was hearing stories about bodies, about negative outsider perceptions, and hearing women talk about their anxieties and their pleasures. If anything, the poler research was a more joyful experience, overall, because of their positivity about the classes and how the classes made them feel, whereas the alternative women[3] were more concerned with how they would, as they aged, manage to balance being alternative enough with being feminine enough; they were, in general, more anxious. The polers did not mention age as something which would limit their lives later, although obviously it would; several mentioned that women of all ages do pole, citing women in their classes of age 60 or above. So what questions did I set out with? I must point out that, from the start, I did not have a fixed interview schedule as such. I wanted to find out why they were attending the classes, what the classes did for them, why they continued going and how the classes fitted into their lives. So I asked very general questions about their leisure time, their exercise history, and, eventually, about their body image, and in this way the answers, more than the questions, found their own way to address the issues. I was seeking 'a rich and nuanced understanding of embodiment; one that puts a premium on the role of social networks in constituting the meanings of the human body' (Waskul and Vannini, 2006, p. 5). I discuss this approach to my interview schedule again in Chapter 1.

In any research the group being studied must have some kind of collective name. Mostly I refer to them as the participants, because they participated in the research and because I consider 'subjects' a problematic term. 'Interviewees' is a little cold although 'respondents' is appropriate when talking about the responses to the online questionnaire. Martha McCaughey (1997, p. xi) chose to call her participants 'self-defensers' because she sees self-defence as both a culture and a movement; she argues that 'self-defensers are inventing new ways to conduct their lives and define themselves as women'. Similarly, I have chosen to call my participants 'polers' because, although pole is not a politicised movement, 'they pole'; are part of a culture and community of polers; and see pole as having improved their lives and how they see themselves – so at points in the following chapters I will refer to them thus. I don't pretend that this is the most elegant, or even the most obvious, name for them as a group but I have noticed its increased use generally within the pole community. Three of the participants suggested it, when I had already begun to consider it, and several others agreed it was a good general collective name, or already referred to themselves as a poler. It would be inaccurate or cumbersome to call them pole dancers or pole exercisers, not least as some of them would prefer one but not the other.

High and low church

What may cause some of the myths and perceptions about pole classes are the different types, some of which draw on lap-dancing techniques while others are squarely placed in gym traditions. I came to think of the differences between classes in terms of high and low church – a slightly bizarre appellation considering the image pole conjures up for some people. What I mean by it is that in the same way that a high (for example, Roman Catholic) church has incense and robes, the 'strippery' classes have high heels and feather boas; and, in the same way that a low church would eschew too many statues and ornamentation, the exercise classes have trainers or jazz shoes (or bare feet), and refute, or at least discourage, comparisons with lap-dancing clubs. I must stress that this distinction was purely for clarification purposes in my own head, and while admittedly overly broad and simplistic, it does provide a quick way of explaining the basic differences of approach (and no offence is meant to anyone who attends any sort of church). The title of this book is *Pole Dancing, Empowerment and Embodiment* but that is so that potential readers are more able to judge its subject matter; elsewhere, I will refer to it as 'pole' so as not to confuse the different approaches of pole dancing with pole exercise. I return to these issues in more depth in Chapter 4.

Figure I.1 Shelly's first 'Superman'. Photograph courtesy of Shelly Glass.

Outline

The outline of the book is as follows: in Chapter 1 I discuss the background of the research. Chapters 2 and 3 aim to provide a context for the empirical chapters which follow; Chapter 2 focuses on the 'mainstreaming' of sex and Chapter 3 focuses on gendered physical activity and exercise. Chapters 4 to 8 provide an overview of pole classes, the students and instructors and the wider pole community. Building on that, Chapters 9 and 10 are two 'case studies': one of a small beauty salon and well-being centre for women in the rural south-west of England which has recently won an award for empowering women through its pole classes and another of all-male classes and what they might tell us about modern masculinity. Finally, in the Conclusion, I consider the preceding chapters and what pole might mean to women's lives.

1
Towards a Feminist Ethnography

> *Narrative inquiry ... can be described as a methodology based upon collecting, analysing, and re-presenting people's stories as told by them. ... Narratives are particularly suitable for portraying how people experience their position in relation to a culture: whether on the margins, in the centre, or on becoming part of a new culture.*
>
> (*Etherington*, 2004, p. 75)

This chapter provides the reader with the story of the research project to contextualise the study but also to provide reflections, and even tentative advice, on the research process. It is not always usual to include a methods chapter in a monograph but I have found that the methodology appendix in *Alternative Femininities* was, and is, being used by so many students that it has convinced me of its validity. I say 'tentative advice' because none of us know what will happen during the research process – Liz Stanley and Sue Wise (1983) advise that it is a mistake to believe that research can be logical and organised and where no problems occur. Having said that, this chapter is not a sequential account of 'how I did the data collection'. But, as Beverley Skeggs (2002, p. 17) argues, 'this chapter provides an underpinning for the rest of the book as methodology underpins all theory. To ignore questions of methodology is to assume that knowledge comes from nowhere allowing knowledge makers to abdicate responsibility for their productions and representations. ... Methodology is itself theory'. This is an assertion echoed by other researchers; for example, Stanley and Wise (1993, p. 164) argue that all research is filtered through the researcher's consciousness and Peter Woods (1999, p. 55) states that 'the account is not purely an objective one that any competent researcher employing the same methods with the same degree of

rigor would produce, but a construction by this particu
Sparkes (1995, p. 280) argues that there is tension betw
researcher) and how people, and events, are represente
valid reasons for addressing the methodological processes.
undertaken for this book was not, and could not be, an e
friendship groups, good (or bad) timing, individual availabi
ness and goodwill, all played a part in the success, or otherw, in the data collection.

In summary, 36 women and 1 man were interviewed in total (23 in the UK, 9 in Sydney and 5 in New York). Of the 37, 16 were students, 18 were instructors and 3 were pole studio managers who also attended pole classes. The age range of interviewees was between 18 and 55 years. I visited a total of 10 studios (6 in the UK, 3 in Sydney and 1 in New York). I was a participant observer for approximately 32 hours in total plus I observed approximately 10 hours in total. For the questionnaire I received 135 responses, from the UK, the US, Belgium, South Africa, Canada, Australia and New Zealand. The age range was an even spread between 18 and 54 years.

Kim Etherington (2004, p. 71) provides her approach to beginning research:

> I need to find ways of working that fit with who I am: my underlying values, my philosophies on life, my views of reality, and my beliefs about how knowledge is known and created. My view of reality or the nature of being or what is (ontology), and my understanding of what it means to know (epistemology) are intertwined. ... My understanding of, and connection with, these concepts guide the decisions I make about my choice of methodologies and methods, and the ways I make sense of the data and represent it. ... I believe reality is socially constructed and subjectively determined.

Etherington's concerns, and her approach, echo my own and so to find out the reality of pole classes I needed to experience them. The first thing about being a participant observer is that one must also engage with one's own participation; 'the issue becomes not so much distance, objectivity and neutrality as closeness, subjectivity and engagement' (Tedlock, 2005, p. 467). It will not always be the case that the researcher will identify with, or agree with, their interviewees although a whole canon of, particularly feminist, methodological literature has grown up around this preference. For example, Donna Luff (1999) interviewed right-wing women and while she experienced moments of rapport with

...em, which disturbed her, overall she did not share their opinions. Of course, rapport is preferable if only because the venture is then more fruitful than a stilted interview would be; and because as a feminist, and a late returner to higher education who now accepts that she will always feel like some kind of an interloper, I feel a responsibility to the women I interview who generously take time to share their experiences (I return to this below). My own reactions to pole classes surprised me because neither did I expect to enjoy them so much, being a non-exerciser, nor to find it so physically liberating and challenging. It reminded me very much of the sort of freedom I enjoyed as a tomboyish girl, climbing trees and swinging on a rope over a stream (more on this in Chapter 4). It made me aware of, and reflect upon, the limitations of my own life as a non-sporty adult female. I have a pole at home now and of the women I interviewed, most bought a pole and had it fitted at home despite the cost and the space required; all the teachers had their own poles at home. So it should not have surprised me that so many women were enthusiastic about talking to me about an aspect of their lives for which they had great passion; one might best describe it as an obsession for some of the participants, as I found out when I approached 'alternative' women. Nor should I have been surprised that some of the polers I contacted were wary and wanted to know what the book was for, who I was, what my opinion of pole classes was and who would read the book. I answered a lot of questions before I started asking any of my own. But, despite having conducted over a hundred semi-structured, detailed, time-consuming (on average 90 minutes long) interviews in my research career to date, including interviews with elderly South Asian women, with grandfathers and with women footballers, I was surprised by both responses. I was touched and delighted by the first response because it is always humbling to meet a complete stranger who then discloses personal things to you and generously gives up their time. I was also anxious about the second response, not anxious that they would refuse to see me but, rather, anxious that they would understand that I hoped to write a positive examination of pole classes; to provide a thoughtful account and answer questions of my own about pole classes. Unfortunately, the conclusion will reveal that I created as many questions as I answered, which is often the case in qualitative research.

Researching pole classes

Another reaction I was surprised by was the continued attention given to my research by some of my male colleagues but for this type of research it

she wanted to understand it (the experience ended up sound-
n illuminating and traumatic):

> was never fully a participant/observer until the night I stripped. 'es, I could observe but being female I could never be a participant/observer with the rest of the spectator/customers. ... I do not reflexively consider myself as simply a stripper, but a researcher who is stripping for her first and only time ... I felt sexually objectified and debased. Moreover, I felt complicitous in my debasement and, by implication, complicitous in other women's debasement by engaging in the activity. ... I felt myself to be abject and I abjectified myself. In short, I got what I came for. I got what I came with.
>
> (Lockford, 2004, p. 88)

Actually visiting clubs and meeting strippers undoubtedly has an effect on the perspective of the researcher as it gives a face, a voice and a position to what was previously 'Other'. Skeggs (2002, p. 19) notes that 'one tradition in anthropology has been to deny to cultural "others" the self consciousness so valued in the theorist. ... During the research I was continually aware of the ease with which those researched can be constructed as objects of knowledge without agency and volition'. Skeggs's research was about working-class women and her participant observation spanned over a decade, far longer than the observation undertaken for my research, but similar in the assumption of a lack of agency in the participants. As Teela Sanders (2004, 2005) and others have eloquently argued, sex workers are not always helpless victims who lack agency; and as both Lockford and Roach found out, and argued, exotic dancers are not just undereducated unfortunates who need to have their consciousness raised by kindly middle-class academic women. 'I learned that sex workers and their supporters voice their concerns through a variety of activist organisations' (Lockford, 2004, p. 186). Similarly, the polers who were professional dancers and performers, organised themselves in 2008 to become members of Equity which afforded more rights around pay, conditions and insurance. But here again we bump against an ingrained misconception which I repeatedly encountered: pole classes are not part of the sex industry and are not (generally, in my experience) in any way linked to exotic dancing in clubs. Of the 37 polers I interviewed, two instructors were ex-exotic dancers which is where they had learnt their pole skills; another had always danced and had learnt to use the pole because she had heard that working in lap-dancing clubs was lucrative; only one instructor was currently a full-time exotic dancer but this was part of a raft of modelling and dancing jobs – I return to this in the next chapter. Mostly the instructors just had a history of dance and exercise. There is no 'standard' poler: students of pole who I met included a lawyer, a plumber and a policewoman, and some of the respondents of the questionnaire included a nurse, a radio journalist, a full-time mum, a phlebotomist, a software developer, an analytical chemist, a barmaid (her own term), a prison drug-treatment worker and a member of HM Forces. In general, their involvement with pole classes was strictly on an exercise, leisure and pleasure basis.

Ethics covers a broad spectrum of behaviours for the researcher, from the obvious which is to be honest about the research, to ensure confidentiality, to obtain institutional approval and so on, to the less obvious which is no less integral but only less 'visible' because these are more deeply embedded. For example, I maintain contact with many participants for the duration of the research and I include denying consciousness to your participants as unethical behaviour.

> It is absolutely crucial that a feminist account of the popularity of G-strings, glossy magazines, cosmetic surgery or any other practice should listen to and treat respectfully women's accounts of their experiences of such practices. This is axiomatic to feminist research. Yet surely this 'respect' does not mean treating those accounts as if they are the only stories that can be told? The role of the feminist intellectual must involve more than listening, and then saying 'I see'. Respectful listening is the beginning, not the end, of the process and our job is surely to contextualize these stories, to situate them, to look at their patterns and variability, to examine their silences and exclusions, and, above all, to locate them in a wider context.
>
> (Gill, 2007, p. 77)

I agree with Gill's assertions wholeheartedly but this paragraph somehow emits a bat-squeak of anxiety for me because of its faint implication that the participants may be unaware of the context, or the silences, and if so, there is a real danger that we listen, go away and inflict our own understandings on the account someone has just articulated for us. Nor should we claim that these are the only stories but surely, while dealing with these stories, their meanings should be of the utmost importance to us. But perhaps any analysis of data necessitates a sort of muffling of the original narratives as Skeggs (2002, p. 15) points out:

> The saddest part of writing this book has been the impossibility of adequately conveying the complexity, resilience, good humour and

> sharpness of the women of the research. They are so much more interesting and insightful than this book can convey in words; sadly, the affectivity of the research has nearly disappeared through the academic analytical filtering process.

Skeggs's words continue to resonate with me and, I suspect, with most researchers at some point: for example, how could I adequately describe the loneliness, longing and rage experienced, as described to me, by a group of elderly South Asian women on arrival in the UK? Their bravery and stubbornness vanished during the 'analytical filtering process' just as their shared humour during an afternoon of tearing up coriander with them also evaporated in the analysis. Finding this balance will always be an issue in qualitative, and particularly in feminist, research; and pole classes would seem to be a very good example of a subject which provides apparently opposing elements: we find exercise, leisure, agency, empowerment, fitness, enjoyment, but also, to many people, objectification, oppression, sex for sale, danger, and lack of agency.

Stage 1: Classes and interviews in the UK

Stage 1 of the research was carried out in 2005 and entailed participant observation of 14 two-hour classes and interviews in Leeds and London. Leeds and London were chosen as they provided appropriate contrast with each other: a capital city where pole exercise was established and a provincial city where pole exercise was just burgeoning. The first part of the study was carried out via participant observation at pole-dancing classes in Leeds, West Yorkshire, to provide descriptive data about women attending and teaching the classes. In total I attended 14 two-hour classes which were run by Pole Stars in Leeds who advertise pole exercise as 'fun strenuous fashionable fitness' (www.polestars.net). The first course I attended was in the spring with a friend (see Holland and Attwood, 2009) and the second was on my own in the autumn. As with any participant observation in the initial stage of data collection, the collection of observations for field notes was time-consuming. Each class could comprise up to 24 women, although numbers were much smaller than that in all the classes I attended, and one or two teachers who were also professional pole dancers often with dance and/or exercise training. Participant observation was chosen for two reasons: first that observers are not allowed into classes; and second, that participant observation is of value to the researcher who is seeking to collect descriptive data and first-hand experience, helping the researcher formulate her questions for the next stage of data collection. In other words, I needed to know what classes were like before I could ask questions about them and formulate theory. There are some debates about the veracity of a project based on observation, not least whether the participants would have behaved as they did if they had not been observed. But as Robson (2002, p. 311) points out it is possible for the 'observed' to be unaware of the 'observer' simply because they are 'so accustomed to the presence of the observer'. The type of participant observation I undertook was as a member of the class and hence unstructured and unobtrusive, so although I made it clear to other members of the class why I was there, the information did not seem to impact on how they behaved or treated me. Certainly participating in something like a pole class is far more enlightening than merely observing or, worse, only talking to people who had done it without having any clear idea of how it is experienced. I was seen as 'one of the girls' because we struggled or triumphed together. However, some more question marks around participant observation as a method are raised later in this chapter.

The second part of the first stage of the research involved semi-structured qualitative interviews with 15 women (six teachers and nine students of pole dancing) between the ages of 20 and 44 years, which were carried out at places to suit the participants in both Leeds and London, after I had completed the first course of lessons. Clearly the subject required sensitivity and I was conscious of participants' reactions to questions about body image, sexuality, 'race' and so on. The interviews were taped with permission and transcribed. They were analysed thematically. Of the participants 11 women were white, two were South Asian and two were black British (their own terms). I met the students mostly at their own homes, or at cafes or bars that they recommended and which were quiet enough to provide adequate recording conditions. I met the teachers mostly at the studio or venue where their classes were held. Pseudonyms are used for most of the participants and for stage 1 of the research they each chose their own, except Rachel who requested that I use her real name. Pseudonyms are something I have considered before and were particularly illuminating in a previous study because of how the names chosen contrasted with other assertions about lifestyle choices (see Holland, 2004). However, again the participants chose pseudonyms because they were their middle names, or a sister's name, in one case a cat's name, and sometimes because it was just a name they liked. This is an issue I return to later. The participants were (instructors) Rachel (her real name) of Zebra

Queen, Shona, Chrissy, Sara, Silke and Annie; and (students) Carrie, Tij, Grace, Kate, Ruth, Carole, Kosa, Jane and Lizzie. In several interviews the issues of time, money and motivation were key, which resulted in a separate paper (Holland, 2009) in which three of these participants were renamed, with their knowledge, 'Amanda', 'Parveen' and 'Sharon'.

The questions asked differed slightly between the students and the instructors but focused on the same general themes which were why choose pole classes, changing fashions in fitness, and their own feelings about body image, empowerment and pleasure. The interviews explored the motivations for choosing to attend and asked what the appeal and benefits of the classes were. I also asked about issues of voluntary erotic female display ('how does the class make you feel?' or 'what do you wear for classes?' usually prompted a discussion about 'feeling sexy'). The interview data not only provided information about the women's experiences of pole-dancing classes but also uncovered detailed and rich accounts of the participants' exercise histories, accounts such as how they had conducted a love–hate relationship with exercise throughout their lives, resisting exercise while feeling that they should and, therefore, how they had sought a less 'exercise' sort of exercise.

Stage 2: Classes and interviews internationally

While 2006 was spent doing other research, by 2007 it became apparent to me that pole classes were becoming even more popular and were mutating and changing, primarily due to increased contact between polers internationally via the Internet. Stage 2 of the research (2007/8) entailed further, but more limited, participation and observation at pole studios in London, Wiltshire, Sydney and New York, and a total of 22 further interviews with polers in those locations. The locations were chosen to provide contrast with each other and to provide an international element to the data especially now that, two years later, pole has become a global (and ever expanding) phenomenon and a much more active and cohesive community, producing DVDs, clothing, publications, calendars and so on. This will be discussed further in Chapter 8.

Again, several of the instructors requested that I use their real names and they are Bobbi of Bobbi's Pole Studio, Jennifer of Studio Verve, Alison of Pole People, KT of Vertical Dance, Genevieve of The Flying Studio, Suzie Q of iPole, James of Aussie Pole Boys, Elena at the Pole Dancing School,[1] Sam of Pole Stars and later of Blush, and Jess of Holistica. If an instructor specifically asked me to include a link to their website I agreed to do so and those links can be found in Appendix 2.

The rest of the participants of this stage of data collection have been given or have chosen pseudonyms which are (instructors) Megan and Evie and (students) Libby, Gidget, Tia, Lilia, Ruby, Keira, Janice, Cindy, Charlotte and Keisha. As with stage 1 I met students at their own homes or at cafes or venues they recommended, and I met the instructors at their studios or at a mutually convenient place (for example, in a café). All interviews, for both stages, lasted about an hour, sometimes more, rarely less, and were recorded and then transcribed.

Stages 1 and 2

There are some significant differences and similarities between the locations of data collection. One of the main developments between stage 1 and stage 2 of the research is the participants' attitude to the name. Pole dancing has now, for those who prefer their pole classes more as exercise than 'strippery', become just 'pole'. This umbrella term is useful when discussing pole classes in general and I discuss the two main schools of thought in Chapter 4.

I cannot include a list of my interview questions as I didn't have one; they remained fluid but always started with a particular question 'what is your exercise history?' and continued from there guided mostly by what the participant wished to talk about. Kim Etherington (2004, p. 56) agonises over how to 'captur[e] the conversations as they had happened and to make make as transparent as possible how I had influenced and shaped the way the stories were told by their discursive nature' but of course the interviewer does guide the subject matter. I had prompts, in case we veered wildly off track, for example, 'how does the class make you feel?' or 'what do you wear for classes?' usually prompted a discussion about 'feeling sexy' or feeling fitter. Participants did report an accrued sense of positive image as a result of attending pole classes. As Shulamit Reinharz (1992, pp. 18–19) argues, 'open-ended interview research explores people's views of reality and allows the researcher to generate theory ... it maximizes discovery and description. ... Interviewing offers researchers access to people's ideas, thoughts, and memories in their own words rather than in the words of the researcher'. Interviewing is not a conversation; it has the form of a conversation but does not follow conversational rules. I have written elsewhere (Holland, 2004, pp. 184–92) about feminist research and ideas about interviewing participants (for example, see Oakley, 1981; Finch, 1984; Ribbens, 1989) so I do not wish to reiterate them here. Suffice it to say, interviewing style does change and develop as time goes on, if only simply because we become more skilled in preparing for and carrying out an interview and it is interesting to go back and look at an old interview 'schedule' if only to exclaim over how differently one would do it now.

Online questionnaire

The online questionnaire (see Appendix 1 for the list of questions asked) provided a list of open and closed questions, multiple choice answers and free text areas in an effort to capture a broad picture of pole classes. It was created online at SurveyMonkey.com which provided a user-friendly way not only for constructing the questionnaire but also for respondents to complete and return it. It was sent to nine pole schools and then forwarded by recipients to various relevant message boards and individuals. Partly its purpose was to provide a way of calculating statistics, for example, there were 132 female and 3 male respondents; 30.4% of the respondents were single and 70.4% had no children. These figures alone can offer us a sketch of who attends pole classes – in this case we see that many leisure or physical activities are easier if you are single and/or do not have childcare responsibilities, a situation apparently unchanged since Rosemary Deem's work in the 1980s.

The questions were designed after conducting stage 1 of the research which illuminated what the main issues for pole classes were, and included questions such as asking about age, ethnic group, occupation, exercise history, body image and so on. The questionnaire was always meant to supplement, or complement, the 'main' body of data from the interviews; to provide a wider lens through which to view, even to check the stories coming from the narrower lens of the interviews. There were 27 questions, which is quite long for a questionnaire and rather longer than I had wanted. This may also account for the numbers who replied to it; had it been 12 questions the completion rate may have been higher. The problem, of course, would then have been that I would not have addressed all my questions: a circular problem. However, many of the questionnaires had sections which were completed at length and so were useable to complement the interview data. Respondents to the questionnaire are denoted by the number of their response (out of 135), their sex and age and their location. So for example, the 12th respondent, a 35-year-old woman from Cardiff in Wales would be identified thus: 12/F35/UK. The questionnaires were also helpful in establishing contact with three further interview participants.

Field notes

Data analyses for interview transcripts, questionnaire data and field notes were carried out in the same way, that is, by repeated study of the material looking for repetitions, patterns and inconsistencies emerging from what may initially seem like a mountain of information, into what Woods (1999, p. 37) calls 'an organised structure', or a series of them. I favour an acceptance of the ambiguities, the contradictions and the tensions, and an embracing of the complexities; sometimes we learn more by looking at the shadows rather than the light. 'The analysis ... values the messiness, depth and texture of lived experience' (Etherington, 2004, p. 81). For all observations and after most interviews I made notes in a field diary, either during the class or interview, or immediately afterwards. Sometimes the notes did not really add to my overall memory and so added nothing to later analysis and reflection, but sometimes I could hear the participant, or I could hear and smell the studio when I reread my notes. There are a cluster of entries where I agonise over what the participants think this book will be like compared to what I know the reality will be. For example, 'again that nagging feeling that whatever they think the book will be, it won't be' and a week later:

> I felt quite cynical today in that I now know I know how to encourage a participant to open up even though I am a perfect stranger, and this used to worry me as if I was pretending to be their friend when I am not. It doesn't worry me any more, I never say I am your friend and actually they never think so. I am friendly, respectful and honest, I cannot do more. My main worry is that it is an academic book and so their idea of what the book will be like may be skewed – but I must let it be, I don't see how I can explain. [I do not include dates with these entries as it would then be obvious to the participant who I had met on those occasions.]

I felt that I was somehow unable to articulate properly to participants that they may find the book to be quite a dry account rather than a coffee table book, which frustrated me intermittently throughout the data collection. In retrospect, I believe these worries were borne of anxieties around the use of people's time and my emotional responses to their stories, and to their generosity and openness; I know that many researchers leave an interview and burst into tears, even when the discussion hasn't

> 'I can feel myself fighting myself' – this prompts more serious discussion about a move, advice, instant encouragement, and a barrage of suggestions, 'put your head down', 'that's why it's good to do a static V first, to see if you have enough strength to hold yourself up there'.
>
> The level of balance, strength and confidence at this level is awe-inspiring. But noticeable that at all the classes, whatever level, I never notice anyone mentioning cellulite or weight, only discussion about how to achieve a move. Few here have noticeable bruises on their arms or legs unlike at lower levels when insufficient strength. Only one woman in this class is in her 20s, the rest in their 30s, the woman who is stiffest is mid-40s and has been doing pole a year but already advanced. 'G' thinks it is a money issue, that women in their 30s onwards have more money to spend on pole classes (which remain expensive).
> [19 March 2008]

I include this long entry because it makes several key points about participant observation, for example, as O'Reilly (2005, p. 85) points out the term is both 'a concept and an oxymoron' – by this she means 'it is not really a method on its own: it involves making notes, asking questions, doing interviews, collecting data, drawing up lists, constructing databases, being active in research. It is never simply a matter of participating and observing ... [and it is] a problematic term, which is interpreted in different ways by different researchers' (ibid., p. 101). So on the day that the long entry above was made, I was welcomed and ostensibly ignored, but I was also an observer, who may have influenced how they talked that day. Afterwards their instructor assured me it had seemed entirely normal to her. That day, because I was sitting on a pink sofa some distance from the poles making notes actually during the class I was not a participant, I was an observer who was collecting data. I wrote fewer field notes towards the end than at the beginning, but I remembered more details for writing up afterwards than I had when I first started. I see this development as indicating that I was so much more familiar with terms, with classes, and with the issues as time went on: 'the odd becomes familiar, the strange usual' (O'Reilly, 2005, p. 100).

A note on writing

During the writing of this book I have had difficulties, with 'block', if it exists. Woods (1999, p. 25) notes that 'the point where rich data, careful analysis and lofty ideas meets the iron discipline of writing is one of the great problem areas of qualitative research'. Ultimately, I realised that the only way I was ever going to finish it was by writing in my own voice. Everyone who reads or writes has opinions about writing styles and about how to write (see, for example, some of my favourites: Brande, 1996; Woods, 1999; Strunk and White, 2000; Truss, 2003; Atwood, 2003; King, 2000). I have a particular writing 'voice' which is recognisable to people who know me or who have read things I have written. I do realise that this 'voice' may offend some people who prefer a more formal style – I apologise in advance if this is the case for you. As time passes I find that I lean more towards a way of writing which tells a story, wherein the voice of the writer (or researcher, or storyteller) is not muted, and I move further away from a drier recounting of academese where the researcher distances themselves slightly from their material. (But neither am I very good at the formal academic style of writing.) I maintain that being at a distance from your material, as we established at the start of this chapter, means you can take slightly less responsibility for it; just as writing in a stiff style-by-rote is a way to mask deficiencies in writing style. As Woods (1999, p. 63) argues, 'we might wish to convey a sense of atmosphere, ethos, mood or tone. We might want to represent feelings and emotions, to re-create people's experiences, to transport the reader to a scene in order to deepen understanding'. Academic writing often overlooks its own possibilities: in the haste to lay out the facts there is no effort to look at the beauty of those facts; even within work such as this, which uses theoretical arguments and ethnographic evidence to explore a particular gendered issue we can find exquisite details about people's lives. As Pelias (2004, p. 11) reminds us, 'a researcher who, instead of hiding behind the illusion of objectivity brings himself [or herself] forwards in the belief that an emotionally vulnerable, linguistically evocative, and sensuously poetic voice can place us close to the subjects [*sic*] that we study'. Participant observation is a particularly appropriate tool for placing the researcher closer, physically and sympathetically, with the participants. However, closeness in and of itself is not enough; what we should strive for above all else is clarity; as Woods (1999, p. 63) also notes, 'clarity and conciseness [are] the main ingredients of the traditional approach'. Both Martha McCaughey (1997) and Kate Fox (2004) argue for academic books to be written more clearly and accessibly, something I myself argued for elsewhere (Holland, 2004). Martha McCaughey (1997, p. x) addresses this difficulty:

> Anytime one attempts to write a grounded, serious book about a wide-spread political and cultural phenomenon in an engaged scholarly fashion, one faces the inevitable dilemma of deciding for whom one is writing. ... Feminist philosophers and theorists have

> challenged central threads of modern thought. Unfortunately, however, their books have appeared remote from the concerns (and vernacular) of everyday women.

While it is clear that many academics take this argument very seriously, it is, unfortunately, still not a common nor a stated aim and many academic writers do not write well, choosing instead to attempt to cover this fact with impenetrable jargon and tortured sentences. Instead, we must endeavour to be understood – it is all about use of language, about the cadence of prose. Kathy Charmaz's (2005, p. 455) advice is, perhaps, the best: 'Write for yourself first. Show, don't tell your reader. Provide concrete evidence.' More than ever I want this book to be read by the 'interested intelligent general audience' because this book is by and about a particular group and I want them to want to read it. Obviously, it is an academic book and so parts of it will be superfluous for a lay reader and possibly even for an academic. I have never read an academic book that wasn't flawed so I certainly don't profess that this book will be without its faults; all its failings are my responsibility alone and it remains to be seen whether I will ever achieve my goals of clarity and conciseness.

Nonetheless this is a book written for two distinct audiences and I am, of course, aware of the difficulties this poses, and also wish to celebrate the fact that this book will have interested readers (even if they disagree with my conclusions) outside academia as well as within it. Alexandra Howson (2005, p. ix) relates an anecdote about Gloria Steinem asking for local activists to be invited to a feminist conference she was speaking at and being told by the organisers that it was impossible because they wouldn't understand the texts being discussed. Steinem retorted 'but they are the texts'. The 36 women and 1 man who I interviewed for this book in the UK, Australia and the US, and the 132 women and 3 men who spent time completing my online questionnaire, are, indeed, the 'texts' without which (or whom) this research would not exist. They were the experts who I went to, spoke to, listened to, and who gave up their time, their experiences and their opinions; many had fully thought-out philosophies which they willingly shared and which shed crucial light on this entirely modern, gendered, sexualised phenomena. 'It is when different audiences are introduced and respond that challenges over the legitimacy of knowledge are produced' (Skeggs, 2002, p. 19). On that note, the next chapter begins to discuss the context of the research.

2
From Circus and Sex ...

> *I don't think I ever associated – I know this is crazy – pole with being dirty or stripper-ish, I don't know why, probably other people do and I still think it's a legitimate way of some people making their money.*
>
> *(Gidget)*

This chapter begins to place pole classes in a wider context, that of what Feona Attwood (2009) and others have called 'the mainstreaming of sex'; Brian McNair (2002) argues that 'striptease culture' now pervades all aspects of Western society and culture from television and films, to art and music, health education and fashion. He sees this as a positive development which reflects the advancement of feminism, gay rights and other elements of society which have previously been viewed as 'deviant'. Academics have turned their attentions to 'striptease culture'; for example, there have been studies about sex work (Sanders, 2004, 2005 on street prostitution; Arthurs, 2006 on feminism and sex work); several North American writers examine the work of exotic dancers (for example, see Kempadoo, 1998; Liepe-Levinson, 2001); there are British studies focusing on the marketing of businesses in the sex industry (Malina and Schmidt, 1997; Jones et al., 2003), or the clients of a lap-dancing club (Frank, 2003), and more recently, a keen interest in burlesque which I discuss further later in this chapter. Having said that, this chapter is not a 'literature review' as such but more of an attempt to provide a snapshot or overview of this particular moment, and the next chapter continues the job that this chapter begins; both chapters seek to provide grounding and meaning for the empirical chapters that follow. The spread of pole classes has occurred within a wider cultural shift. As other authors have argued, the sexualisation of culture has permeated our lives, from

Ann Summers' shops and strip clubs 'emerging in a ferment' (Connell, 2005, p. 92) on the high street to pole dancing being featured on the *Oprah Winfrey Show*, to men's magazines which contain much of the same content as porn magazines (for example, see McNair, 2002; Attwood, 2005). More recently erotic/soft-porn magazines for women, such as *Scarlet*, have been published, which are the offspring of glossy post-feminist women's magazines (for example, see Macdonald, 1995; Whelehan, 2000).

As McNair (2002, p. 25) argues, many women have grown up with issues seemingly resolved, 'particularly those younger generations brought up in a world where ideas which had once been the province of radical feminism were increasingly viewed as self-evident common sense'. Thus these young women might see porn as relevant to their sex lives, rather than in automatic opposition to them as women. Similarly, Ariel Levy (2006) argues that the new 'bawdy' society is not going to render women sexually liberated but, indeed, points only to how far feminism has yet to go – that rather than making any real advancements, women have instead 'joined in' with their own objectification, becoming misogynistic, but failing to be ironic or funny and failing to benefit their lives or that of other women. Ariel Levy (2006, p. 99) argues that the new 'Female Chauvinist Pig' or FCP adopts 'a passion for raunch' simply to attract men: men have proved they like girly-girls but FCPs have proved they want to be like men and have expressed disdain for women. So they somehow have to find a way to show 'men' that they are both like them and can appreciate watching other women, at the same time intimating that they too have sexy underwear on under their powersuit despite their aggression and their binge drinking. This 'girly-girl' figure has reached her nadir in the popular imagination. '[It] has become the term women use to describe exactly who they do not want to be: a prissy sissy' (ibid., p. 101). But pole cannot be described as girly-girly, because of the strength it requires, nor as butch either, because of its roots in lap dancing, but also because in many classes women actually do 'dress up' (to which I return in later chapters). Pole should not be classified as bawdy simply because of its positive physical benefits which do not rely on sexualised display, although it can co-opt sexualised display if required by the individual poler.

'Sweeping statements about pornography's relentless objectification of women or its embodiment of patriarchal structures of dominance and submission can not be borne out by any detailed examination of the many different types of pornography that exist' (Attwood, 2002, p. 94). Certainly, there is a wide range of images and even soft-porn images such as those seen in *Nuts* magazine, which can indicate a startlingly old-fashioned modern sexism as Attwood (2005) has argued. But this is to assume that all heterosexual males respond in the same way to violent images, and by no means is all porn violent, as Connell (2005) has also pointed out. As Valerie Steele (1996, p. 187) notes, 'many people believe that pornography causes perversion and sexual violence. But this is like saying that country-and-western music causes adultery and alcoholism'. As McNair (2002) has argued, feminist work challenging pornography has resulted in feminist arguments about choice being used to defend pornography and in women using pornographic images for pleasure. In the same way classes based on a physical activity which has hitherto been only for the entertainment of men is now being marketed as being a positive, woman-friendly, women-only leisure class because of advances made by feminism. Emma Rich (2005, p. 495) has argued that feminism occupies a shifting and ambiguous place in the lives of young women today, where they both distance themselves from the subject position of feminism but also utilise its ideals and the gains that it has achieved for women; 'positioning themselves in often complex ways, against and within the discourses of feminism and gender' (ibid., p. 496), which I return to in Chapter 4. McNair (2002, p. 24) quotes Segal pointing out that the positive aspects of female sexualities are continually submerged beneath assertions that women can only be the victim:

> [Segal] wrote that 'while the voices recognised by mainstream culture as "feminist" remain those busy demonising male lust and pornography as a metaphor for evil, it will not be possible to find any confident – or even hopeful – popular affirmation of a feminist sexual politics'.

Websites such as Suicide Girls (http://suicidegirls.com/) do attempt to present a confident and empowered version of pornography, celebrating 'alternative' beauty via the photographs sent in by women with tattoos, body piercings and so on. But, as Segal notes, every attempt made by Suicide Girls to assert that their choices are about an affirmation of a feminist and political approach to sexuality, will be refuted by feminists who are not sex-positive and can imagine porn to be only about male lust and violence. Similarly, Feona Attwood (2006, p. 89) argues that modern approaches to love and sex contribute to the confused position that phenomena such as pole classes occupy:

> [S]exuality therefore takes on a very particular set of qualities as it assumes such a central, yet nebulous role in articulating our bodies

> and our pleasures and in making our claims to individuality, to a self for itself, to our status in the world, to our embodiment for others and for sex itself. These qualities mark out a quite specific sensibility of sex which is linked to the broader conditions of our social world. ... Given this, it is hardly surprising that sex is able to signify so much, and to signify so contradictorily both the personal and the political conditions of our existence.

Pole classes, then, fit into this model: they are/have been about sex, they are also about embodiment, pleasures and individuality – and represent both the personal (exercising for fun) and the political (women's continued place in the sex industry). Pole classes continue to be burdened with a kind of schizophrenic image which has, just over the course of this research, settled down into a more athletic norm. For example, an early sort of home pole was the Kitten Pole and one of the accessories available was a fake smoke-alarm to fit over the fitting in the ceiling – in this way you could take your pole down if you had visitors and even disguise the fitting. Later, these vanished and with the new X-Pole (easy to install with no drilling) women tend to keep their poles up on a daily basis if they have the room. The Pole Stars website describes the classes firmly as women-only, positive and supportive, and claims that

> we have a policy of inclusion which means everyone should feel comfortable. ... Pole dancing is a great way to stay in shape, but there are loads of hidden benefits such as feeling sexy, building confidence and self esteem and creating a *supportive female only* environment [my emphasis].
>
> (www.polestars.net)

This, then, shows how pole classes have always had, even among themselves, two paradoxical positions: one, good, clean, energetic fun where you meet other women and increase your self-esteem and on the other hand, something to hide from your friends or family. It is perhaps because of these kind of contradictions that others argue that pole dancing (and pole classes) exists, and exists in the mainstream, because of negative cultural developments which undermine women's lives (on a broad, social scale) and which, more specifically, threaten advances made by feminism which have improved individual women's lives.

Strippers and exotic dancers

As R. Danielle Egan et al. (2006, p. xxii) argue, 'exotic dance is, according to anthropologist and dance scholar Judith Hanna, a "lightning rod" for cultural conflict over bodily display and over when, where, and for whom such display constitutes an acceptable form of self expression'. Exotic dance has become more visible as culture becomes more sexualised. As Jones et al. (2003, p. 215) point out, 'while traditional striptease clubs have long been seen to have a rather "seedy" image, and often a back street location, the new generation of lap dancing clubs claim to offer a sophisticated experience and with an annual turnover estimated to be in excess of £300 million, they are one of the fastest growing elements in the UK's leisure services industry'. Strippers and exotic dancers have attracted academic interest (for example, Schweitzer, 2000; Wesely, 2003; Sloan and Wahab, 2004; Bott, 2006; Rambo et al., 2006; and Downs et al., 2006). By the same token, stripper memoirs have been published such as *Bare: On Women, Dancing, Sex and Power* (2002) in which Elisabeth Eaves reminisces about her time in strip clubs. As Lesa Lockford (2004, p. 185) notes:

> Before someone challenged me to consider that the women who work in the sex trade might not consider themselves to be as disempowered as [some feminist authors] argue, I accepted this feminist view. ... It wasn't until after I began interviewing dancers and going to the clubs that I became aware of the literature that had been published by sex workers. Indeed, much of that literature has been published in an effort to temper feminists and conservatives who tend to pathologize sex workers in their efforts to make their respective rhetorical and political arguments against pornography.

Similarly, edited by three ex-strippers who are now feminist academics, *Flesh for Fantasy* (Egan et al., 2006) is a collection of auto-ethnographic chapters which reflect on the place of the body in embodiment; not the theoretical body but the actual, lived, leaky, sexual body. 'The efforts among third wave feminists to destigmatize sex work are often perceived as antagonistic to the goals of second wave feminism. However, the self-reflexive writings of third wave strippers do not necessarily oppose second wave feminism; rather, they pick up on the sex positive and sex radical positions within the second wave that have too often been obscured by mainstream media and movement spokespersons' (Egan et al., 2006, p. xv).

Bobbi's feelings of being in control and having choice echo Egan's assertions about the self-reflexivity of many strippers:

> I think until you actually do it, you can't really talk about it personally, I think it is, I'm completely conscious – I don't do drugs, I haven't done anything like that in my life, I'm completely in control of everything I choose to do and stripping was something that came so naturally to me, I can't explain it and not in a 'I'm hot shit' way. I love expressing myself sexually, I feel so good about it and I'm smiling, they're all smiling, everybody is happy and nobody loses, I don't walk out of there going 'oh my god, I feel so used and deprived', yes I walk out and go like 'that was great, I did a really good show, did the splits and pulled that off well', that's how I feel when I walk out and I think the most successful strippers do it the same way. They go I did a bad show [or] I did a good show, not 'oh I feel that guy was giving me a bad look and I got a dirty vibe', we don't take it like that, we either perform well or you have a bad day.
>
> (Bobbi)

Bobbi expressed her love for stripping which led to the opening of her pole studio:

> I had my love, my passion which was stripping in the clubs. ... Love it, I can't tell you! And I positively think because of that, because it's in my blood to do this [pole studio], is why businesses close, because I actually love it and I can promote the positive side because I do it myself, I'm not just some outsider saying 'it's good for health and fun and fitness because I read it in [a magazine]', I'm not like that. I'm like 'I do it for a living, this is how it feels, this is how I look and you can take my word for it' kind of thing.

Which was echoed by Lockford's (2004, p. 86–7) findings:

> If I once had an essentialized notion about women who strip, I found as I encountered sex-positive feminism embodied by the sex workers I met that I had to abandon those conceptions, however discomforting abandoning them might be. ... Yes, I met and interviewed women who had nothing but bad things to say about their experiences as exotic dancers and whose viewpoints and experiences support mainstream feminist arguments against sex work. Yet I also met women who emphatically tell me they love working in the clubs,

> women say they experience the work as anything but abjectifying, women who are quite resolutely feminists. ... I met women who find stripping empowering for how it teaches them about their bodies, and for how it teaches them about the power of desire, and for how it helps them to love themselves.

I return to the concept of empowerment, and the limitations or possibilities of the term, in later chapters.

From the questionnaire responses two women intended to become lap dancers; one had done research to inform her teaching; and five had been, or were, professional pole dancers in clubs. Two (Bobbi and Suzie Q) instructors do, or have, worked in lap-dancing clubs; three (Silke, Evie and KT) have worked as podium dancers although not in lap-dancing clubs; and three instructors (Rachel, Megan and Jess) did it briefly, or at least visited clubs, as research for their teaching:

> I went and did a couple of nights lap dancing just so that I could tell lap dancers what it was like, because I was teaching them anyway, and I thought 'it is not fair to teach when I don't really know', so I went and did a couple of nights. And no, it doesn't really matter – it is nearly all attitude. It is not dependent on your looks, your body, whatever – your age. There is a demand for every sort of woman there is. I think in a way, if you are a little bit different in one way or another, in a way it is to your advantage. (Rachel)
>
> I did it for a while, a short while, just so I could say 'look, it's like this'. It is quite different to classes, let me tell you and I, well really, I didn't like it much. (Megan)
>
> It is not a different form – it is the same. It is different in the sense ... have you ever been to a pole club? The girls in pole clubs are there to show their bodies. The pole dancing that they show isn't athletic – it is wiggling around a pole stripping. So in that sense it is completely different. But in the other sense it is still a pole and you are still dancing, and you do get some girls who work in clubs just to go on the pole, and they do upside down, inverts and spins and things – so I don't think there is a difference. (Jess)

The distinctions between working as an exotic dancer and hating it, or loving it, or doing it briefly as 'research', or being a dancer who can pole, or being a pole instructor, or being a pole student – all these categories and distinctions, with their crossovers or changes, are often not taken into account. For example,

> [t]he proposition that having the most simplistic, plastic stereotypes of female sexuality constantly reiterated throughout our culture somehow proves that we are sexually liberated and personally empowered has been offered to us, and we have accepted it. But if we think about it, we know this just doesn't make any sense. It's time to stop nodding and smiling uncomfortably as we ignore the crazy feeling in our heads and admit that the emperor has no clothes. ... All the cardio striptease classes in the world aren't going to render us sexually liberated.
>
> (Levy, 2006, pp. 197, 199)

No, of course not and certainly no pole school or studio I went to even attempted to claim such a thing; their aims were quite simple, to enjoy exercise, to boost self-confidence and body image and to encourage others to do so. In fact, Levy's concept of who does pole, and who does stripping, are the most simplistic and plastic stereotypes. Levy absolutely fails to address the key issue, just as some feminist academics have refused to credit exotic dancers with having consciousness and agency. What is that key issue? It is to listen to the voices of the polers themselves and to assume that women can decide for themselves.

New burlesque

Burlesque has, just since the 1990s, had a revival and has attracted as much academic interest as exotic dancing (for example, Willson 2008; Ferreday 2008; Fargo 2008), and burlesque classes and workshops are run in many towns and cities just as pole classes are. But burlesque classes focus on how to shimmy, how to remove your gloves, how to twirl a parasol and how to wear pasties (nipple tassles). Burlesque has a long history, coming out of early Victorian music halls, and was ribald, comic, suggestive and knowing. Its appeal lasted until the 1960s; famous burlesque artists included Gypsy Rose Lee. Its revival used many of its original trappings, such as music, dance, singing, partial stripping and props (chairs, umbrellas, outfits, fans, feathers and so on), but now is performed with a more ironic, post-feminist approach. The most well-known stars include Immodesty Blaize and Dita Von Teese. Burlesque shares many of the same issues as pole does, such as sexualised display, objectification and so on. But pole classes and burlesque are different, for reasons which I will make clear in the later chapters. Only four participants mentioned burlesque and their attitude was that burlesque shouldn't be conflated with pole, that the two are different:

> Burlesque is really big here in Australia at the moment, so people are doing burlesque pole, as in striptease type of thing, striptease pole – they bring it into a lot of other things. (Sam)

> GENEVIEVE: I think burlesque is a bit of a, like you say, side dish.
> SH: It's separate?
> GENEVIEVE: It's different, because you could do burlesque with a chair and an umbrella, or you could do burlesque on a pole which I know some people do. You could do a comedy act like Charlie Chaplin. It doesn't mean that pole dancing and comedy acts are the same. Aerial stuff definitely, like somebody who has never done pole before might struggle with aerial stuff, a lot of aerialists go from pole to aerial.

> Burlesque, my friend does that, it's taken off here. But it's not the same as pole. They might use a pole in their act, some do, some don't. But it's not the same. (Cindy)
>
> I don't imagine burlesque will last like pole dancing will. It is just such a different thing, quite specialized, very theatrical, you know, it is just different. Appeals to different sorts of people too I imagine. (Carole)

Pole, in and of itself, is more athletic than burlesque; burlesque does not require athleticism in the same way that pole does. Obviously a burlesque performer may choose to use a pole as part of her performance (burly-pole). Burlesque and strippery-pole performance both have a routinised performance to music, an outfit and sometimes (as at Bobbi's Pole Studio) a separate pole name for your pole 'persona' – but apart from a routinised performance to music, pole which is about dance and/or fitness will not require a theatrical outfit or a performance persona. Several recent newspaper articles have criticised burlesque as moving away from its saucy, ironic approach. For example, the *Guardian* newspaper (Penny, 2009) recently printed an article in which it stated:

> [Burlesque] has been sold to the public as something subversive, even feminist, a democratic form of objectification which welcomes any woman, regardless of age or dress size. In practice, this seems to add

> up to the less-than-radical notion that women who have cellulite can be sex objects too. ... But stripping of any kind can only offer passive, cringing empowerment at best.
>
> (http://www.guardian.co.uk/lifeandstyle/2009/may/15/burlesque-feminism-proud-galleries)

Again, we see the assumption that exotic dancers cannot be reflexive about their choices. Admittedly though, there will always be those who earn enough to see their subjectivity as empowerment because it earns them enough to do so. As with the pole world, there are differences between the rank and file and the 'celebrities'. For example, on a flight to Barcelona recently I read a copy of *JetAway* magazine (February/March 2009, p. 9) in which was an article called 'Dita Goes Crazy': '"some people say that what I do isn't very liberating" coos the raven-haired pin-up. "I say it's pretty liberating to get $20,000 for ten minutes' work"'. This will obviously contrast sharply with the experiences of most burlesque performers, working the pub, bar or club circuits in their hometowns, and also with even the highest paid, well-known (among the pole community) poler. It also underlines the very subjective and individual effects of 'empowerment', which some feminists despair of. But as Shelley Budgeon (2001, p. 18) argues:

> Individualism privileges the worth of the individual at the expense of the collectivity [or] it can also be a source of agency at the micro-level of everyday practices. ... Whilst the [participants] had no sense of a collective political tradition ... [they] exercised a politicised agency at the micro-level of everyday social relations.

Pole classes do not claim to be about politicising those who attend but, unlike burlesque classes, they do offer improved physical fitness and confidence which, in turn, leads to a sense of agency.

Embodiment and empowerment

Many of the participants mentioned empowerment, as do many of those who have researched pole dancing, lap dancing or stripping; for example, Rambo et al. (2006, p. 217) discuss such difficulties:

> In some radical feminist discourses exotic dancers are passive sex objects who lack agency and unwittingly reinforce traditional patriarchal values. ... If a dancer claims she is not exploited or oppressed, if she expresses job satisfaction or enjoyment, resists oppression, or

> feels like an exploiter or powerful herself, then she is characterized as a victim of false consciousness – a passive agent and cultural dupe who has internalized her oppression.

Lockford (2004, p. ix) states that she began her research because her teacher said to her, 'I bet if you asked women in strip clubs if they felt objectified, they'd say no' to which she agreed but then started to think about how this might be further explored. What is clear is that she didn't expect it to impact her own version of feminism quite so strongly in that she started to see the strippers that she met as intelligent, reflective agents. The same is true of Catherine M. Roach (2007, p. 17): One of the questions which haunts her throughout her research is, if a woman can, after weighing up the pros and cons, decide that stripping is the best job for her, and choose to do it, being able to stop doing it when and if she wants to, why isn't stripping therefore OK for that woman? Why can stripping categorically be no good for all women? She comes to the conclusion that this question has its limits because which of us do freely choose our profession, which of us is truly free of glass ceilings, of sexism and of the expectations of our backgrounds? This is a relevant argument for the polers because a pole class can be good for women in that its physical and emotional benefits cannot be refuted: it can increase their confidence, it can make them fitter, they can get enjoyment and a sense of achievement from it, they can make friends. The women I spoke to were certain that it was a good thing and initially wary of me because they assumed, through their previous experience and a general cultural stigma, that I would be asking the very questions that I was asking and then almost automatically drawing negative conclusions which implied that their choices and their rationalisations were somehow at fault.

> It seems to me ... that feminist theory needs to be less concerned with achieving theoretical closure and more interested in exploring the tensions which the body provokes. This would entail using the tensions evoked by the contradictions ... as a resource for further theoretical reflection. It would mean embracing rather than avoiding those aspects of embodiment which disturb and/or fascinate us as part and parcel of our theories on the body.
>
> (Davis, 1997, p. 15)

Empowerment and embodiment would seem to be at odds with each other in that one is often about transgressing the limitations of gender and the other is often about the parameters of it. Embodiment carries with it issues about regulation and discipline, about resistance, submission, norms and expectations. Shilling's (1993) account of the dialectic between socially shaped embodiment and embodiment beyond the social, provides a useful framework to begin to understand the body. Later feminist theorists focused on lived female embodiment; the labour required to maintain one's own and other bodies (for example, see Dryden, 1999; Beagan et al., 2008). Feminist work around embodiment has not been unproblematic; often, just as in 'mainstream' sociology, the female body has been defined as problematic, deviant, leaky and out of control; 'reduced to sex, defined in turn as neutral platform for the inscription of gender identity' (Howson, 2005, p. 57). Howson also points out that there are notions about the female body as a burden and that Simone de Beauvoir influenced early theorists about the female body as burden rather than celebration (ibid., p. 55). Certainly, there has been ample attention, much of it feminist, paid to, for example, the pregnant (Tyler, 2000, 2001), the breast-feeding (Smyth, 2008) and the menopausal body (Greer, 1999). But as Alexandra Howson notes, 'the body has only recently emerged in the past decade as a theoretical focus for feminism' (Howson, 2005, p. 9) and much existing work around the body is not based on empirical data. As Howson argues,

> contemporary feminist approaches to the body increasingly focus on the generation and reformulation of theoretical frameworks. ... Hence, the starting point for much sociological and feminist commentary about the human body is text rather than matter. ... The starting point matters. The data that form the focus for theorizing the contemporary body are primarily other texts that are used to reformulate, reconsider and revise prior claims about the body. Consequently, the body within both sociology and within feminism has become subject for more abstraction, not less.
>
> (Howson, 2005, p. 9)

The starting point for this study is, as Howson recommends, matter rather than text. 'If the body is something that people do then it is in the doings of people – not their flesh – that the body is embodied; an active process by which the body is literally real(ized) and made meaningful' (Waskul and Vannini, 2006, p. 7). Kay Inckle (2007) draws a distinction between a sociology of the body and embodied sociology, arguing that the former reduces the body to just that, 'the' body, whereas the latter

> engage[s] the messy, complex, and contradictory factors at play in human experience, as well as the essentially emotive, corporeal

> and intersubjective, visceral, sentient nature of our being. As such, embodied sociologists abandon abstract categories and the ideals of analytical tidiness.

Here then, we see that the endeavour is not just about bodies but also about what we learn from bodies and the lives they lead. As McCaughey (1997, pp. xi–xii) points out, 'corporeal feminism insists on examining the ways such ideologies become inscribed and contested at the level of the body. [Self-defensers] cannot kick and scream their way out of systematic oppression' but power can be 'resisted through the lived body'. McCaughey argues that feminists have neglected to consider self-defence training as a way of 'subverting the embodied ethos of rape culture' (ibid., xii). So too are some feminists rather too eager to assume that pole cannot be anything but negative because it is oversexualised. As I have argued elsewhere (Holland, 2004, p. 10), the problem is that women in general come to be seen as 'mindless consumers, in thrall to the power of media images' and so seeming as if they need good, sound feminist intervention to save them from themselves. McCaughey (1997, p. xii) demands that we consider how we might embody alternative ideologies and argues that one good example is self-defence which 'demands that feminism take seriously the corporeality and pleasure of that resistance. It demands that feminism get physical'. I would argue that pole is another example of how women can get physical and are attempting to resist ideologies about what is appropriate – the problem with pole, of course, is that it does not contest sexed and gendered embodiment, at least, not to the casual observer; 'sexed embodiment in which male aggression and female vulnerability are cultural corporeal paradigms' (McCaughey, 1997, p. xv). And yet the improved strength and confidence of polers does contest gendered embodiment. The end result is not, and often cannot be, more than physical strength but can offer women a genuinely improved body image and sense of self. For example, tattoos on women can be seen as 'emblems of empowerment and as badges of self-determination' (Mifflin, 1997, p. ii) and polers undoubtedly see their abilities as indicative of self-determination.

Pole classes

Pole has a place among other sexualised dance and performance forms such as burlesque, Chinese silks and trapeze, but its trajectory is also one of exercise and athleticism, of Nike as well as Dita. Pole dancing is linked to two present-day (but age-old) obsessions (sex and exercise) and yet it retains a romantic and traditional similarity to trapeze and high-wire acts, which may explain part of its appeal to the public imagination. Angela Carter's (1985) character Fevvers in *Nights at the Circus* was both grotesque and fascinating, a trapeze artist with wings, who had started her career in a freak show – much of her popularity stemmed from her mix of physical strength with sexuality which modern pole dancing echoes. Pole dancing began, according to many accounts, in Canadian strip clubs in the 1980s but arguably has its roots (however tenuously) in a number of forms such as Maypole dancing and Mallakhamb (yoga practiced on a wooden pole, mainly practiced in India), and more recently the Cirque du Solei have used Chinese pole acts (http://www.flyingstudio.co.uk/History.html, accessed June 10 2009).

However, pole classes are not a miraculous riposte to hegemonic discourses about the gendered body and women's experience of physicality, mainly because the image of pole remains sexualised to a significant proportion of people. For example, in August 2007, an Italian woman in Tokyo emailed me lamenting the fact that her professors thought pole dancing was too 'trivial and sleazy' to warrant PhD research. In the summer of 2008 a woman in Norway told me that Norway still had no pole classes because of its residual salacious reputation. Its neighbouring countries did have them but Norway still did not, despite its reputation for promoting fitness.[1] Another pole contact, a Malaysian student studying in Cardiff, told me via email in summer 2009 that, because her course had come to an end she 'had to sell my X-pole last week (which felt like giving up a boyfriend or a really good friend) and am already missing my pole lessons. No chance to dance in Malaysia, unfortunately, as my parents don't know that I do it!' Katherine Frank (2006, p. 203) discusses part of Chris Rock's recent addition to his stage act in which he says that 'one of his main tasks as a father ... is to "keep her off the pole". And just to be safe, "my daughter is staying away from all poles, that includes monkey bars" he says. ... As the mother of a baby girl, and as a former stripper, I found that Rock's comments gave me reason to pause'. Despite these lingering attitudes, pole classes do offer alternatives, a way to embody alternative ideologies such as strength and coordination, and sexuality, and this remains an important possibility, one that has its limitations, but one which offers to women the pleasures of resistance. As Anna Aalten (2004, p. 264) points out, 'feminists criticized the representational practices in ballet, because 'women do not represent themselves in ballet' (Adair, 1992, p. 116) and because as an art form ballet is 'rooted in an ideology which denies women their own agency' (Daly, 1987, p. 17)'. Feminist work around ways of resisting feminine embodied norms such as the hour-glass figure or the pneumatic blonde, has shown us that different types of female bodies can be a response to the limitations imposed on women's bodies: for example, eating disorders can be a form of resistance (Bordo, 1993), and body building can also be seen to be resistance (Mansfield and McGinn, 1993; Grogan et al., 2004).

In 2006 and 2007, in the *European Journal of Women's Studies*, a debate raged over three issues in an article which took two seemingly contrary examples of types of dress, headscarves and porno-chic, to illustrate the autonomy and range of girls' clothing choices. The debate then developed into a discussion about the differences between Dutch and English feminism, about choice, respect, autonomy and agency within discourses of 'decency' and discourses of feminism and about participants being treated as objects rather than actors. Although not immediately apparent how these arguments related to my work on pole classes, I began to see a general correlation around the issues of 'decency' and sexuality and echoes of my own concerns that the voices of the polers I had met would be subsumed in an analytical discourse which posited that they could not possibly be in charge of their own agency. These are the issues which I would like to examine here, relating them to pole classes.

Linda Duits and Liesbet Van Zoonen (2006, p. 104) argue that there are 'gendered subtexts in contemporary social tensions about dress', drawing together an array of concerns and anxieties about girls, women, their bodies, their sexualities and what they choose to do with them. The statement could as easily be read substituting the word 'dress' for the words 'pole classes' (indeed, could be replaced by any number of phrases which currently trouble feminism about the lives of women). Indeed, most issues that are screamed out in the media have a gendered subtext, such as the concentration on Rosemary rather than Fred West, on Myra Hindley rather than Ian Brady, and on girls who drink too much alcohol and have fights or casual sex rather than focusing on boys who do exactly the same. Duits and Van Zoonen argue that concerns about what girls wear are subsumed in hegemonic discourses about 'decency', but also in feminist discourses which are similarly damning: the girl cannot win because she is seen to be a mindless dupe of capitalism, or of patriarchy, or of the media, or a slave to her sexuality. The discourses of decency and feminism appear to have been conflated when it comes to the mainstreaming of sex in culture and both discourses seem to suggest

> that there is a 'good', political way to reveal your body, and a 'bad', consumerist way. Finally, both discourses ignore the fact that sexuality is not an essential feature of bodily display and dress, but the effect of the reaction of others.
>
> (Duits and Van Zoonen, 2006, p. 108)

As both Beverley Skeggs (2002) and Valerie Walkerdine (2003) argue, working-class women must guard their 'respectability' to avoid being defined by their sexuality and this is certainly something implied by some of the polers; for example, one of the questions on the questionnaire was 'what are the negative things about pole classes?' to which many responded as this selection illustrates:

> Narrow-minded opinions of others about pole dancing can be quite draining. (2/F34/UK)
>
> The stigma some people still attach to pole dancing. We are not a bunch of sl*gs and people need to get over this and accept it as a form of fitness and dance. (59/F27/UK)
>
> Can be associated with sex industry, sometimes people have a negative reaction. (10/F20/Aus)
>
> Only when dealing with non-polers. I feel sometimes I am censoring myself in fear of what others may think, e.g., employers or elder relatives. (13/F27/USA)
>
> Occasional remarks by idiots. (16/F37/UK)
>
> The way people stereotype pole dancers. (17/F33/Aus)
>
> Having to constantly explain it's not sleazy. (18/F35/UK)
>
> Dealing with absolutely negative opinions from closed-minded people. (6/F47/USA)
>
> My boyfriend doesn't like the idea. (75/F24/UK)
>
> [This is one of three responses where a male partner didn't approve.]
>
> It is a shame that so many people still frown upon the art form. (86/F23/UK)
>
> People still mix up pole dancers and lap dancers. (103/F31/UK)

Replies in the same vein were received from 11 other respondents, using words such as tarty, slutty, sleazy and stigma; I return to these issues in Chapter 7 and in the final chapter. Herein lies the dilemma for me. Second-wave feminism was never the dour, sexless movement its detractors claimed it to be.

> This chasm between what was seen as an 'orthodox' sexless feminism or 'correct' femininity, and a post-feminism or 'antifeminism' of sexualized dissent permeates feminist history and remains a sticking point. ... Feminist resistance against female sexualized display is vehemently grounded in the desire of women to be accepted as thinking individuals and not, as Griselda Pollock stated in 1981, 'explicitly as cunt'.
>
> (Willson, 2008, p. 7)

After all, feminists have always concerned themselves with issues around sexual liberation, sex work, choice, sexualities, embodiment and freedom. But the discourses began to lean towards those where women were in crisis, where women's bodies were vulnerable, such as rape, anorexia, teen motherhood and, latterly, issues such as self-harm and binge drinking. Obviously this has been a strength of feminism and, unfortunately, a necessity. Martha McCaughey (1997, p. 148) has written about women and sexual agency, arguing that the danger with anti-sex feminism is that it does not challenge the stereotypes about women's sexual pleasure and agency, nor does it question or subvert some of the myths around sexual violence and rape:

> For women, sexuality has been both repressive and dangerous as well as pleasurable and exciting. Too much focus on women's sexual victimization ignores women's experience with sexual agency. ... By the late 1970s some feminists began to embrace forms of sexual expression that had been deemed forms of 'false consciousness', for instance, sadomasochistic sex or bisexuality. The anticensorship or pro-sex feminists tried to gain support for feminism not by denying differences among women but by emphasizing them. Thus the pro-sex feminists have urged women to speak as powerfully in favour of sexual pleasure as they do against sexual danger. In doing so, women might break through stereotypes of women as naturally passive, essentially different from men in erotic styles and preferences, and thus challenge some of the very myths that perpetuate sexual violence.
>
> (McCaughey, 1997, p. 148)

The general approach of the exercise polers also implies a striving for respectability and not wishing to be defined by their sexuality but, rather, by their athletic prowess. But if we add into the mix the mainstreaming of sex in popular culture, the position of the neoliberal subject (the individual rather than collective), the aims and gains of both second-wave and third-wave feminisms and 'critical moments' which inspired the women to start pole classes – what is the result? The result is women attempting to find agency through physicality. Agency, as Wendy Parkins (2000, p. 62) has pointed out, is 'necessarily embodied'. She quotes Nancy Fraser's concerns: 'what we often seem to lack is a coherent, integrated, balanced conception of agency, a conception that can accommodate both the power of social constraints and the capacity to act situatedly against them (Fraser, 1992b: 17)' (ibid. 59). Can pole offer a path to Parkins's conception of agency? The following chapters aim to answer that question. This chapter has attempted to place pole classes in its wider context of the mainstreaming of sex in popular culture, and debates about agency, 'decency' and sexuality, all of which I return to in the final chapter. '[Feminist] scholars have regulated, disciplined and controlled female bodies and selves through the discourse of commodification. ... Through the gendered discourses of the commodified body, the narrative possibilities of the dancer are limited. If she persists in this behaviour, scholars will label her as defective. She is not permitted to define herself' (Rambo et al., 2006, p. 221). This book aims to allow polers to define themselves. The next chapter will develop the contextualisation of pole classes by examining more closely the gendering of leisure and exercise and asking where and how pole classes fit into that framework.

3
... to Fitness and Leisure

I wish I had a dollar for every time I heard 'God, this is so much harder than I expected'.

(*Sam*)

This chapter continues the task of contextualising and theorising pole classes, which was begun in Chapter 2. Commonly feminist scholars have focused their attention on women who play team and/or male-dominated sports such as American football (Kotarba and Held, 2006), soccer (Scraton, Caudwell and Holland, 2005), windsurfing (Wheaton, 2004) and Australian rules football (Wedgwood, 2004). Women who play such sports embed their embodiment within the very masculine cultures and often face accusations about their sexuality or a loss of 'femininity'. In contrast, pole is predominantly taught and attended by women, and so poses different questions. Pole lacks credibility because it is not about women trying to succeed in a masculinised sport. Waskul and Vannini (2006, p. 9) argue that skill display in women's American football 'constitutes an alternative to these hegemonic discourses, thus providing women with an oppositional symbolic zone for the redefinition of their bodies'. Instead of issues around lack of access or acceptance, the focus moves to individualised, biographical, cultural, physical accounts and the potential for agency within the limitations of being seen as 'only' a feminised form of exercise – but, just the same, 'in competitive sports, women's bodies are sexualized but men are portrayed as powerful' (Kotarba and Held, 2006, p. 153); we will return to this issue in Chapter 10. Arguably, pole is trivialised and sexualised, even by feminists, because it is 'just' women taught by women and is not about collectivised political activism; the stereotype of a class is that there is some swinging round a pole and some gyrating of hips by women wanting to feel sexy. To some extent, it fits into the 'carnivalesque' tradition described by Feona Attwood (2002, p. 93) with its bawdiness, its association with erotic pleasure, its fascination with a body that can be seen as 'vulgar'. Like Arabic (or 'belly') dancing it has a history (or herstory) of being a dance or performance wholly or primarily performed by women for men – although, obviously, Arabic dancing is a much more ancient form and has folkloric links, not to mention celebrating the female form in larger sizes.[1] We need not consider it seriously as an exercise form or a pleasurable leisure activity because it makes us uncomfortable: it isn't 'serious', it doesn't have any intellectual qualities, the women in the classes wear skimpy clothes and it has its roots in lap-dancing clubs. Yet, apart from the last point, it is just the same as any other exercise class and, as Tara Brabazon (2006, p. 65) argues about aerobics classes, it 'is an important intervention in the masculine modalities of sport, providing a pleasurable site for community building among women'. Similarly, Maxine Leeds Craig and Rita Liberti (2007) found that their participants reported feeling more comfortable and confident because they had chosen to join a women-only gym. (I return to these points in Chapters 8 and 9.) As Anna Aalten (2004, p. 272) argues about ballet,

> [i]t is obvious from the stories of the dancers that ballet offers women an opportunity to excel physically in ways that are comparable to sports. In ballet, women can find an opportunity to excel physically, but without the association with masculinity.

Pole, too, is a physical activity which gives women the chance to learn to excel physically without associating the participants with a masculine sport and the problems that come along with it (Cockburn and Clarke, 2002; Caudwell, 2006; Rich, 2005). On the other hand, it is still seen to be 'inappropriate' because it is also performed in strip clubs. As Lesa Lockford (2004, p. 27) has pointed out, 'the power through which women have been defined in terms of their bodies is palpably experienced by women through the material effects and social practices constraining their agency'.

Exercise (her)stories

All women's activities, physical or cultural, take place within a complex network of sexism, racism, heterosexism, capitalism and body consciousness, often body dismorphia; although women are physically expected to be, in general, timid there is also agency, pleasure, and dominance. According to the online questi (53 people) said they had always loved exercise usually hated exercise and 35.4% (46) said they off. Many of the participants of my study said that th classes because they had become bored with other sorts of fit pilates or aerobics, or that they didn't exercise as such but occasio sought out 'something to do' because they felt they should. Yet if pole classes were truly about fitness, the classes, in the early stages, failed to deliver a good workout to its students. When I began this research in 2005 many schools ran classes where only two poles were shared by up to 12 students, making the purchase of a home pole necessary to show any benefits. Nowadays it is rare to have to share a pole in a class which has greatly increased its potential to provide an adequate workout. Pole classes fit with the norm of women being perceived as more suited for non-combative sports, but it is not without competitiveness, which I return to in Chapter 8.

Heléne Thomsson's (1999) research about the history of exercise in the lives of 50 Swedish women illuminates, very helpfully, some of the prevalent attitudes and tensions. Thomsson's participants, aged between 20 and 60 years, demonstrated that although they felt they should exercise because of pressures about how they should look, they were also subject to constraints about what they thought was appropriate (for example, sweating or getting red in the face being unfeminine). As Thomsson (1999, p. 36) argues, 'the best way to reach an understanding of the complex phenomenon of women's participation or rejection of exercise is through … women's own narratives'. These narratives operate within several discourses about gender. 'A woman's body is seen as a tool for the construction of femininity, and many of women's body-related beauty practices (which may include exercise) are effects of the normalizing regime' (ibid., p. 39). But within these discourses many women will never feel happy with their bodies, always feeling that they are too fat or too thin, too short or tall, or too old. Exercise, then, becomes just another way that they fail and which, paradoxically, exacerbates their already negative body image. Segar et al.'s (2006) study examined women's motivations for exercising and their findings show that women with 'body-shape motives', such as toning or weight loss, were significantly less physically active than those with 'non-body-shape motives', such as fun or sociability (Segar et al., 2006, p. 175). However, as Thomsson (1999, p. 39) points out, 'shaping oneself as a woman, or doing gender, can be seen as a leisure activity [in itself] for ly women' and pole classes do, at least on first glance, function perfectly as a place to do gender and shape oneself as a woman within very specific hegemonic parameters. It is only when we look closer that we realise that pole classes also have transformative potential, and, most surprisingly, potential to improve the body image, sociability and confidence of women who do them. But, as I examined elsewhere (Holland, 2009), the classes also inspired the women to overcome various barriers to attendance: namely, time, money, motivation and tiredness. These are the key barriers which women continue to face and overcoming them becomes, in and of itself, part of the pleasure of attending; the very act of achieving attendance, coupled with exercise, then become ways that they don't fail.

Most of the instructors had a background of exercise, for example, Jennifer, Bobbi, James and Elena are all professionally trained dancers. Both KT and Suzie Q were active as children:

> Yes, I was a massive, massively sporty child – suffering from dyslexia, I failed at everything at school academically – but what I was, I was really good at running, so I used to run for my County, I used to play netball for East of England, I used to represent my team school so, all the way through ... netball, hockey, absolutely everything, just excelled at sport and loved dancing, loved doing everything you know, just really, really active. ... I was very pushed by my parents, you know, my mum played netball, my mum was very sporty. (KT)
>
> When I was a kid I did gymnastics for quite a few years, I did that competitively, I have always been very into sport, always playing hockey or touch football or rugby, you know, anything. ... It wasn't actually till I started teaching [pole] that I started taking it seriously and actually getting quite good at it. ... Because when you are teaching sort of 10 or so classes a week you are getting a lot stronger, which means you can do other stuff. (Suzie Q)

Several pole students did have a history of regular exercise, for example:

> I did a lot of gymnastics from the age of 7 every day and dancing school, tap, ballet, acrobats the whole bit and sort of did that right up until about 18. Then did nothing for a while because I had it drummed by my parents that dancing will never take you anywhere, so you've got to go out and get a real job. (Gidget)
>
> I've done Greek dancing from when we were young so that wasn't my form of exercise, I was skiing and fencing, I used to travel around

> with the fencing and when I married and had children I took the girls, my daughters to fencing and then they stopped and then I thought I need some sort of exercise, so I did salsa dancing. ... Fencing was when I was in my early 20s and then continued to into my 30s. ... Because I used to love Errol Flynn and you know I always wanted to be a pirate and swing from the chandeliers and get my sword and fence on the boat. ... I did skiing as exercise as well. ... I did do the salsa dancing at the end of my 40s and early 50s. ... but I didn't like the teacher so I stopped. (Tia)

Of the six teachers from stage 1, two had no dance and/or fitness background:

> I don't do exercise and I am a rubbish dancer. ... I never thought I would end up in any sort of career that had anything to do with dancing. I am very lazy – I don't exercise, but I do like dancing, so it is one of the few physical things that I do because I don't do exercise. ... I am accidentally fit. (Rachel)
>
> I sort of fell into the pole dancing thing and the lessons, doing, teaching the lessons, I just fell into it cos I had done some of the classes myself and I just picked it up real quick but I am not a keep-fit type. (Annie)

All the students from stage 1 of the research claimed to have a history of limited or sporadic physical activity, or to be mostly sedentary or even total exercise-phobes. Carrie and Tij's responses were typical; 'I hate exercise, to be honest' and 'I hated P. E. at school, I hate gyms, I hate it all, it's mind-numbing, I can't be arsed'. However, both of them enjoyed going to nightclubs to dance and I return to the appeal of dancing later in this chapter. Many of the questionnaire respondents replied that they usually hated exercise (two examples are 28/F33/UK and 33/F31/NZ). From stage 2, three instructors said they did not have a history of enjoying 'organised' sport or exercise:

> I was like the most unfit person you ever came across – but with teaching like 3 times a week, I was getting so much stronger, and my upper body definition and all the rest of it – and loving it. It was so much fun. (Sam)
>
> I don't think I've done anything I've classed as exercise but it probably was, for example, running around [the supermarket]. ... I've never kind of gone for an exercise class that is restrictive in any way

> and doesn't make you express yourself and is regimented. ... I don't like fitness classes, I find them very negative. (Genevieve)
>
> Not happy about exercising – before I started pole [which] all just fell into place real quick, but before that, I wasn't what you would call someone who exercised, I walked about, that was about it. (Megan)

Walking is a popular way of exercising without exercising; and, like dancing, it fits into the 'princess and the pea' approach to physical activity, which I explain later. Lilia is 47 and a lawyer with 4 children. She described to me how she cautiously approached her decision to begin classes, taking into account her age and her ambivalence about organised exercise:

> I have no dance history whatsoever, I didn't do any physical exercise until I was about 19 or 20 and then I started doing aerobics and basically did that until I had kids [age 31] ... and then I have just walked everywhere and that is the only exercise I have ever done. ... I put on a pair of comfy shoes and I walk to work so I am [still] a good walker.

Walking is advocated as very successful in maintaining fitness and weight loss (for example, see Gilson et al., 2007) and several participants, including Genevieve, Sam, Janice, Cindy, Charlotte and Keisha, mentioned that the type of activity they had most frequently engaged in before pole classes was walking. So Lilia's decision to try pole, considering her age and lack of organised exercise experience, could have been disappointing for her:

> I hadn't been to a gym in a thousand years, in the window it says there was this pole class so I read the brochure and found it really interesting but I didn't do anything about it right then. ... I thought 'oh that sounds really good' and because it was in the gym it took away the connation of it being sleazy and my recollection [of it being on the Oprah show] was that it makes you really fit and if something can make you fit and is fun it must be really interesting. ... I still sort of was like 'oh I don't know if I can do this, it is going to be all these young girls and I am this old chick'. Then I thought 'oh why not, I will give it a go'. ... I went to the very first class feeling anxious and ... there is this real mix [of women] so it looks interesting and then I met Jen and she was just wonderful.

While Lilia's initial experience of pole class was positive, thanks to the women who attended and the instructor, it was her ongoing responses to the classes which ensured her continued attendance – sociability, enjoyment, achievement rather than failure and no anxieties about being perceived as masculinised. In these ways pole hides itself as something other, and more, than exercise.

The princess and the pea, or, exercise in disguise

In the fairy tale about the princess and the pea, a young girl is tested as to whether she is truly a princess. The form of the test is whether she can feel a pea which is placed under her mattress. After several mornings of waking up covered in bruises, and an increasing number of mattresses being placed between her and the offending pea, the girl is finally pronounced a true princess. In this analogy I see 'sport' or 'P. E.' as the pea and any participant who had previously disliked and avoided exercise, as the princess. Pole, then, is the pea that the princess couldn't feel and it is interesting to consider why it has such success at attracting, and retaining, previous sporadic or non-exercisers. As Belinda Wheaton (2004, p. 1) argues, 'lifestyle [or] alternative sports [such as surfing or snowboarding] ... are different to the western traditional activities that constitute "mainstream" sport' and points out that 'many commentators are agreed in seeing such activities as having presented an "alternative" and potential challenge to traditional ways of "seeing," "doing" and understanding sport. ... Many have characteristics that are different from the traditional rule-bound, competitive and masculinised dominant sport cultures' (ibid., p. 3). Pole fits into this remit which may explain its success at evading the easily-bruised princess (or serial non-exerciser). 'For some time, it has been recognised that PE and sport have a history celebrating hegemonic competitive masculinity, featuring white middle-class practices and values' (Rich, 2005, p. 503) which, as Wheaton argues, can be resisted through alternative types of sport or exercise.[2]

So why have participants such as Genevieve, Tij and Sam previously avoided organised physical exercise? Feminist researchers such as Scraton (1992) and Wright and Dewar (1997) have argued that physical education classes are exclusionary to girls and the hangover from experiences of P. E. at school can last a lifetime, as many of my participants proved. As Adair (1992, p. 41) notes, 'passivity is a common characteristic of women's physical experience; the still groups in the playground, restricted movement of women on the street wearing tight clothes and the almost stationary positions on the factory floor'. Acceptably 'lady-like' bodily posture is static and tidy, taking up as little room as possible. Thomsson's (1999, p. 52) participants avoided exercise because they didn't want to appear sweaty, red or ugly revealing a fear of being seen to be unlady-like or unfeminine. Her study found that they rationalised their exercise and non-exercising phases of life with various discourses, for example, 'if' (if my children ... or if my husband ...), lack of willpower, time constraints and fear of being selfish (ibid., pp. 47–8). Many of the non-exercising participants of my study named some or all of these reasons for why they had previously exercised sporadically. So some of the main issues which continue to preclude women's access to leisure time include time due to family and work commitments (Deem, 1986; Shaw, 1992, Miller and Brown, 2005); childcare and housework (Shaw, 1997; Dryden, 1999); notions of what is gender-appropriate (Wedgwood, 2004; Grogan et al., 2004; Cooky and McDonald, 2005); and body image (Grogan, 1999; James, 2000; Segar et al., 2006). Another barrier was simply boredom, or fear of boredom, which was connected to a fear of failure – if they were bored they would ultimately fail by not continuing, or if they failed they would become bored and, again, not continue. The appeal of pole is because it is so new, it doesn't have connotations of P. E. or other sorts of exercise at which they feel they have previously 'failed' – instead it is more akin to tomboyish, out-of-school childhood activities which they did enjoy.

Additionally, many of the students chose pole exercise because of its high media profile and the benefits (enjoyment, toning up) that magazines or celebrities attributed to it: Darryl Hannah, Kate Moss and Madonna were just some of the famous women who have said that they use a pole at home – primarily in 2005 and 2006. In 2007, Pamela Anderson said (in *The Sun* newspaper online) that she had a pole in her bedroom at home but that her two sons used it as much as she did, and Natalie Portman enthused about the benefits of pole after learning to pole for a film role. As Rachel said:

> With some of them I think it is a fashion. They think 'Kate Moss does it, so I am going to go and give it a go because the posh celebrities do it'. It is nice to have an item of celebrity lifestyle.

Several participants (including Silke, Chrissy, Annie and Charlotte) had first encountered poles in nightclubs or bars or at hen parties where part of the day was spent at a pole session. Some had come to pole classes because they had surprised themselves by enjoying another sort of dance, most commonly Arabic (belly) dance or salsa, which they had begun because it had been 'exercise in disguise'. An exercise form called 'dance' appears to have more appeal to non-exercising women because of its connotations of creativity, glamour and gracefulness. Dance appears to appeal to non-exercising women because it can offer more possibilities for feelings of abandon or liberation – despite the fact that, contrarily, formal dance training would be hard work within rigid parameters – although Lilia notes that her lack of dance training inhibits her ability to learn whole routines:

LILIA: I can't dance, I still can't dance.
SH: Yes, you can dance now, can't you?
LILIA: I can't dance, everyone does the routine and I shuffle along but then it doesn't bother me.
SH: Are you more comfortable with tricks than with linking moves?
LILIA: Yes, yes.
SH: And why do you think that is, because there is no dance background?
LILIA: Yes, it is a language I can't quite read, and I just don't sort of get the movement, and I am still quite tight. ... There is the one routine I actually like and I can really get into it but yes, I am not a dancer, it doesn't come naturally, but I still do it and it just takes me longer to learn.[3]

Worryingly, this has a faint shadow of 'failure' about it except that Lilia doesn't perceive her inability to dance as boring or failure but enjoys the class anyway. As Thomsson (1999, p. 44) points out, many women express their distancing from sport as a way of being women; sport is a key outlet culturally for the demonstration and perpetuation of hegemonic masculinities. But it is acceptable to dance; dance, despite its requirement of discipline and athleticism, is feminised. So even women who have always hated exercise can be lured in by exercise that comes disguised as something else, exercise that might offer more fun or exercise, or that can be seen almost to be not-exercise. Instructors and students alike were aware of this element of pole's appeal, for example:

> It is a great way for women who haven't exercised, suddenly they are there once or twice a week working out, and enjoying it. I think [pole] is fantastic for them because yes, some women are intimidated by the gym. (Suzie Q)

SH: Do you find that some of the girls that come, say in general, how many of them have got exercise backgrounds and how many of them say, god I've always hated exercise but I love this.

BOBBI: Yes, I don't know everyone, but I've heard equal comments from both sides. A lot of 'I don't do anything at all, I need to find something that interests me so I can get fit' and then a lot of 'I go to the gym all the time and I'm over it, I need an alternative'. ... They get hooked, because it does work and it's, I think the difference between this and any other sport is you're not looking at your watch, and thinking, 'okay 5 minutes to go and then I can get off', [instead] it's'oh no, it's almost finished' and before you know it, you've burnt more calories than a lot of other things and met new friends along the way.

> Pole isn't really like exercise to them. It's hard work but we don't notice that part. (Megan)
>
> It is exercising without realising it. (28/F33/UK)
> So much fun you don't realise you're having a work-out. (33/F31/NZ)
> You don't realise you're exercising. (101/F28/UK)

In this way pole appeals to non-exercisers and exercisers alike and has another weapon in its armoury which is 'do what's right for you', first mentioned in Chapter 1.

'Do what's right for you'

For stage 1 of the research, the teacher of the first course of classes that I attended was a professional pole dancer with exercise-instructor training and a full-time university student. The classes were designed by the company who ran them and the teachers were required to adhere closely to the plan given to them. Classes began with a warm-up session followed by mat-work, then there was pole work that lasted over an hour and finally ended with cool-down exercises. The mat-work was rigorous with exercises for the abdominals and upper body in particular. For all the students this was the least enjoyable and most difficult part and many of them resented how it intruded on the class. For example, Kosa said,

> [s]he can be a bitch about it, like a P. E. teacher, I had tears standing in my eyes and she was going 'a bit more Kosa, a bit more', I felt like telling her to stuff it.

and Jane said,

> I don't enjoy the sit-ups and that as much as being on the pole, I know she said we have to build up our muscles so we can do the

> pole stuff better but I hate all that, I have always hated all that stuff.

The mat-work was, for most of the students, too much like exercise that wasn't in disguise. Pole classes offer the opportunity for physicality, pleasure and feelings of empowerment – unexpected results considering its history – that is, for exercise wrapped up in other more enjoyable things. But, like the princess and the pea, no amount of 'padding' could disguise that mat-work was exercise; for that reason most of the participants hated it and it was the one period of any class where we were most uncooperative. Wright and Dewar (1997, p. 80) describe their own 'intensely pleasurable experiences' in relation to physical activity. This was not the experience of many of the participants of this study and so they resented the mat-work which resonated too strongly with the sort of exercise they had always historically associated with pain or failure or boredom rather than pleasure. Like Thomsson's (1999, p. 43) participants, they associated certain sorts of activities with discomfort. It wasn't the stationary mat-work that they were there to do; if they wanted to do sit-ups (which they didn't) they could do them at any time; they were there to swing around a pole. It wasn't just mat-work which came as a surprise either. Some students or instructors enjoyed linking moves (such as head or body rolls, hip gyrations, or rolling on the floor and so on) whereas others (including myself) did not enjoy them:

> I normally do 2 hour classes here and we do warm up and stretching, wriggly bits, bits of dance steps and then pole moves – I don't hold with teaching people to fiddle with their bits and lying around on the floor too much. ... Yes, well basically, as people are leaving I say 'practice wriggling around on your bedroom floor – I haven't got time for that here'. (Rachel)

SH: I wasn't ready for the bits in the middle, the linking moves. ... I went with my friend Feona and we were just like 'uh-oh', we weren't ready for that; [the instructor] would say 'be sexy' and we would go 'oh no'.

ALISON: Well, 'sexy' is a word banned in Pole People classes ... you can say elegant, you can say anything, sexy is just a horrible word. It makes people panic. Absolutely panic. ... It's the only thing that is banned, I am fairly open-minded! But it makes people freeze up. It is such a loaded thing, it goes so deep for so many women, you know? ... You are trying

> to create a positive environment for people to feel free, you know, happy with their bodies, and you don't want people to feel they have to be a certain way – if they want to do it like that, fine, if they don't want to, that is fine as well.

> At first I didn't like it. By the end I loved it, I had no idea I could get my leg behind my head! (Grace)

'Do what's right for you' works in pole classes because not only can women go to 'high' or 'low' classes according to what suits them but within that they can also negotiate with their instructor about how much they feel able to do. An example of that came in both Lilia's and Lizzie's interviews:

> LILIA: I am not very flexible and there are moves I don't think I will ever do.
>
> SH: Can you do the splits?
>
> LILIA: No, I can't even touch my toes. ... Because I am not flexible Jen will always give me an option for me to use so I don't feel like I can't do things. ... No one ever makes you feel that you can't do that, and they always give you options. ... I will do a move and of course their legs go up to their ears and I say 'well it is not about getting your legs high or wide it is about getting the right line and you can do that in any option', they taught me that, so every one of us is supportive no matter what shape we are and no matter how flexible we are.

> I can't do the splits, I just can't, we have tried and tried, it would put my hip out or something, and the, I mean, in the end she said 'ok we can work around that, the splits aren't what you do', which we laughed but I felt, you know, a bit – anyway, so we look now at the lines I make, at the extensions I make with, you know, arms, legs, how I stretch my neck, and ultimately I don't think, we think you can just work around it. (Lizzie)

Rachel commented that sometimes people want to go at a faster pace:

> Beginners generally can't hold themselves up very well – they are not used to holding their own body weight, so it is basic moves. If you keep going with the basic moves, you build your strength up to be able to do the next set of moves. ... But if people have got the bottle [nerve] to do it – if, on their first class, they want to try chucking

> themselves upside down, I will show them how to do it in the safest possible way. And if they do it, then bloody great.

And, similarly, 'do what's right for you' might not be about what you can or can't do, it might be about finding your own pace, as Gidget remarked:

> I didn't think there would be anything untoward, I never thought that. It was all about me though, like I walked through that door and I didn't really care how [instructor] was, it was like, am I going to be able to do this, am I going to look stupid, that was my biggest thing. Am I going to get there and the other girls are going to 'huh, look at that, doesn't that look stupid', that was my biggest thought. ... But in the first couple of lessons when I used to do things [instructor] would come round and you know, she would tell us how to do something and she would walk around the class and have a look at how we were going and say 'oh goodness, you've done ballet, you've got lovely turned out legs' and I thought, 'oh I'm still turning my legs out' and you know, so you think 'oh, I can sort of do this, this is ok'.
>
> (Gidget)

Gidget's anxieties about doing what was right for her were less about what she couldn't do – as a trained gymnast and dancer it was unlikely that the splits would pose a problem for her – but more about being accepted by her teacher and being allowed to go at a pace which suited her during her first underconfident months at classes. Doing what is right for you, and being given options to suit your abilities, are, in general, not how the non-exercising women remember their previous experiences of P. E. and/or fitness classes. Add to this the novelty of pole, the sociability of classes and the physical benefits and we begin to see pole's appeal and why it propels women to begin to exercise regularly after years of not doing so. Finally in this chapter I consider reasons why the 'do what's right for you' guidance was ever necessary in the first place.

Swinging like a girl?

Some commentators refute the idea that pole can be positive (for example, Levy, 2006 and Siegel, 2007). As Martha McCaughey (1997, p. 163) argues,

> [f]eminists have suggested that women's physical activities, even if seemingly empowering, still connect their self-worth to body management. Feminists have worked against the shaping of women's

> bodies to the figurative ideal, and want to have 'real' bodies. The problem is that all 'real' bodies are to some extent imaginary constructions. ... Without a disciplined body project, one cannot be effectively subversive or conformist. In other words, systems of corporeal production per se are not oppressive, as many feminists appear to assume.

In their responses to pole classes, detractors, whether they be feminist academics or journalists needing a quick fix of titillation for their readers, mirror knee-jerk responses similar to that of popular media coverage of sportswomen: the assumption being that pole, since done primarily by women, can only be about women's objectification and sexualisation whereas all the women I spoke to talked about fun, friendship and increased confidence and strength. Heléne Thomsson (1999, p. 39) argues that women are caught in a double-bind: 'the body is both the site of women's entrapment and the vehicle for the expression of themselves' necessitating a constant balancing act which is subject to the scrutiny of others. Although bodies can, with limitations, convey alternative ideologies (Mifflin, 1997; McCaughey, 1997; Holland, 2004), the body in an exercise class is also 'a negotiated self, a site of both anxiety and display' (Brabazon, 2006, p. 67) and the pole body attracts anxiety and even derision. As McCaughey argues, corporeal feminism insists on examining the ways that ideologies become both inscribed and contested at the level of the body, and pole exercise is both interesting and difficult because it is about inscribed ideologies of certain sorts of hegemonic femininities and also about ways to contest those hegemonies, such as muscle building, physical confidence and creativity. Obviously, women cannot 'kick and scream their way out of systematic oppression' (or, in this case, climb, invert and swing) but 'power can be resisted through the lived body' (McCaughey, 1997, pp. xi–xii). The confidence and ability to really engage with physical activity is still commonly 'trained out' of women from adolescence onwards as feminist theorists have argued (for example, Lees, 1993; Sharp, 1994; Ussher, 1997). Young (1990, p. 146) points out that

> [n]ot only is there a typical style of throwing like a girl, but there is a more or less typical style of running like a girl, climbing like a girl, swinging like a girl, hitting like a girl. They have in common first that the whole body is put into fluid and directed motion ... and second that the woman's motion tends not to reach, extend, lean, stretch, and follow through in the direction of her intention.

Young's thesis resonates with the early stages of the pole-dancing classes and echoes the arguments that exercise almost sets women up to fail, being as it is a 'holy grail' which is never at an end. But, as Young (1990, p. 147) also concedes,

> [t]here is no inherent, mysterious connection between these sorts of typical comportments [above] and being a female person. Many of them result ... from lack of practice of using the body and performing tasks.

We see this in some of the interviews:

> I'm still pretty rubbish, to be honest with you, it doesn't come naturally to me. (Carole)
>
> Some people have got the dancey thing and they are used to moving their bodies and they don't feel awkward doing the wiggly silly bits and stuff like that. (Rachel)

CARRIE: I never remember to point my toes, I never remember to extend my arm fully, I just can't seem to, you know, I just can't keep it all in my head at once.
SH: But you can drive?
CARRIE: Er, yes, uh-huh.
SH: You keep a lot in your head at once to be able to drive, right? You change gear, read the road, check the mirrors, all that. ... So ... ?
CARRIE: I get it yes, I do, I suppose with exercise I am still somehow not, it doesn't somehow feel right to stretch like that, to be so ...
SH: Physical?
CARRIE: [laughs] Yes, physical.

Carole's and Carrie's assertions that physical activity remains 'unnatural' or 'not right' echoes Aalten's (1997, p. 49) work about the training female ballet dancers are given so that they 'generally make smaller movements, showing less strength than male dancers. Their leaps are primarily supposed to be supple and ornamental or tiny and rapid. ... [She is told] not to make such strong movements with her arms'. But as Young (1990, p. 155) points out, a woman is damned if she does and damned if she doesn't; many women simply are not used to being physically expansive, and certainly it is not universally accepted that they can or should be. Women rarely sit with their knees wide apart. 'A woman's distance from her body and physical hesitancy may come from seeing herself as a sex object for others. And yet, she must keep herself "closed in" because to use her body freely is to "invite" sexual objectification (not to mention claims of crassness). Thus keeping in your enclosed space is a defense against being leered at, touched and accused of inviting rape. Through self-defense instruction, women lose a certain bodily comportment of femininity' (McCaughey, 1997, p. 120). This was the defining feature of the pole classes: women who had previously disliked exercise (for many of the same reasons as Thomsson's participants) or who started exercise but continually dropped out (just as some of Segar et al.'s participants), arrived at the classes enthusiastic but resigned to failure. They were out of practice of using their bodies. They were, as the weeks went by, the same women who gave the 'physical tasks [their] best effort [and] were greatly surprised indeed at what [their] bodies could accomplish' (Young, 1990, p. 147) and, as they did so, both lost and gained 'bodily comportment of femininity' – lost it because they were getting stronger and more physical, climbing, swinging, stretching, inverting; and gained it, because they were pointing their toes and not bulking up and were becoming proficient at a safely feminised/sexualised activity. As Jennifer explains:

> There is really technical aspect to it, if you can see it, but as a dancer I could see that, so that is what addicted me to it. And the strength aspect. And it rekindled my dance again in a different way and one thing led to another and then I fell into a situation where I was teaching and then at that point started researching and doing my own, the way that I saw pole at that point was it is very static, you do a move and then you do a body roll and then you do a move and then you do a body roll ... so it became challenge to me as to how I could incorporate elements of real dance. Dance with movements on the pole and create something that looks more lyrical than what you know and kind of takes it away from that just strip type exotic dance movements. So that has been something that I have been like pursuing for the last 6 years now.

Pole classes, then, can be seen to shift away from only having connotations of being exercise for strippers or Barbie dolls, or as reinforcing negative body image and feelings of failure, to becoming a site for exercise-phobic women to relearn how to use their bodies, how to set themselves physical tasks and how to achieve them. No wonder that the participants spoke of empowerment; of feeling physically empowered by the classes. Wright and Dewar (1997, p. 91) found that many of the women in their study used the same definition for empowerment in their discussion of the pleasures and pains of exercise, and the feelings of achievement, and certainly this fits in with the accounts of pole classes. Women's sporting achievements are almost always conflated with their gender and sexuality and men's with power and strength (Schultz, 2004; Kotarba and Held, 2006) which automatically erodes any potential for empowerment. For example, Brandi Chastain's winning goal in the 1999 World Cup was ultimately overshadowed by the now iconic image of her lifting her shirt in celebration – the sports bra she revealed, and not the winning goal she scored, excited all the attention, as it would not have in the men's game (Schultz, 2004, p. 186). But as Wedgwood (2004, p. 141) asks, 'what then of subversive forms of female embodiment, such as powerful, skilful, and/or assertive forms?' Pole, despite claims to the contrary, is arguably subversive, powerful, skilful and assertive – in the context of classes, at least. And so the next chapter begins to examine pole classes in more detail.

4
What is a Pole Class?

It is the most loving, feminine, supportive environment that I have ever come across.

(*Lilia*)

Most people have some idea of what they think pole dancing involves, certainly they have a picture in their heads about how it must look. The most common image is of a young, nubile, slim woman, moving seductively from a standing to a squatting position, with the pole between her legs. The slight variation of this image is of the same woman hanging from a vertical pole in a seemingly impossible gravity-defying position, while wearing little or nothing – this latter image is less common although it takes more skill. Muscular endurance and coordination are aspects of pole dancing least likely to occur to people; as are balance, hours of practice and feats of daring. I did a double take on reading Anna Aalten's (2004, p. 267) description of ballet because it could so easily have been written about pole; replace 'ballet' for 'pole' and the whole paragraph still makes perfect sense:

> Openness, verticality and stylization are the basic aesthetic principles of ballet. The beauty of ballet is created by the straight lines of the extended human body going outward and upward and by the artificiality of the movements. Ballet dancers produce a spectacle in which upward-aspiring straight lines and an illusion of weightlessness are central elements. But human bodies do not consist of straight lines and they are inevitably subjected to the law of gravity.
>
> (Aalten, 2004, p. 267)

Being a proficient poler involves strength and a certain amount of recklessness, and creates, in Aalten's words, a 'spectacle [of] ... upward-aspiring straight lines and an illusion of weightlessness'. But, as Aalten (ibid., p. 268) also points out, 'the difficulty of the technique and the need to mould the body in a particular form make the daily training process necessary'. Pole is hard work, as I discussed in the previous chapter and as most of the participants pointed out:

> Yes, it is an art, it's a learning thing, yes definitely. With aerobics you can go and switch off and still do the movements, whereas with pole if you switch off you have lost it. (Jess)

SH: So what is the difference between intermediate and beginners? Just more technical?

RACHEL: Yes, like you have got your basic right way up swings where you have got 2 hands a leg on the pole say, and then you have got your things where you are just holding on and using your arm strength. And you have got upside down ones where you are holding on with 2 hands and then you have got upside down ones where you are not holding with any hands, and then you have got ones where you are holding with your hands but not any legs, and there are lots and lots of different possibilities.

TIA: [At advanced level] they have to exercise every day, they have to do stretches, they have to move every single day to keep up to that level – I don't know, Gidget would have told you she exercises.

SH: Every morning!

TIA: You see, she's a gymnast.

Pole classes are still relatively new which may explain the confusion or hostility about them, both reactions connected to pole dancing in lap-dancing clubs. This chapter aims to shed some light on the classes.

The genesis of pole classes

The history of pole dancing is mentioned in Chapter 2. According to Genevieve Moody, on her website, it

> originat[ed] in the strip clubs of Canada in the 1980s. ... The origins of Mallakhamb (meaning 'man of power' or 'gymnasts pole') which is yoga practiced on a wooden pole and on rope (mainly practiced in

Figure 4.1 Beginners' class. Photographer: Caroline Fearn. Copyright: Polepeople Dance and Fitness Ltd.

> India), may date back to 12th century but the discipline is suspected to be formed as recently as 250 years ago. A second discipline directly relating to todays pole dance known as Mallastambha (meaning 'gymnasts pillar') using a mainly iron pole was used by wrestlers to build up strength. Mallastambha seems to be no longer practiced. Today pole Mallakhamb is practiced by men and boys and rope Mallakamb (aka Indian rope trick) is practiced by women and girls.
> (http://www.flyingstudio.co.uk/History.html, accessed on 10 June 2009).

It was interesting to me that there is no definitive, agreed-upon starting point for something which has become as prevalent as pole dancing – and, latterly, pole classes. I saw this as illustrating a certain anarchical element to pole: while talking to polers there were occasions when I realised that I was thinking about the way in which self-professed Riot Grrls of the 1990s talked about their zines, bands and approach to both modern feminism and life in general (for example, see Leonard, 1997; Garrison, 2000; Armstrong, 2008). There were very similar feelings of being on the frontiers of something new, something female-focused and/or feminist, something which had the potential to give pleasure, or, ideally, even to enhance lives – and all that with an undertow of defiance and independence. Pole classes have never been in the hands of any sort of dance or fitness 'establishment' but have always been initiated and run by women driven by a belief in pole's potential; pole is not owned by companies such as *Nike* or *Fitness First*, nor is there even a particular person or group credited with being the leader or one true initiator. Pole answers only to the pole community; obviously, within that, as I note below, there are divisions and differences of opinion. Most of the instructors claim to be among the first to have thought of classes, or to have pioneered a type of class; it is impossible to know who was really the 'first'. For example, Rachel started Zebra Queen in 2000; Genevieve opened The Flying Studio (initially Strictly Pole Dancing) in 2002; Alison started Pole People and Bobbi started Bobbi's Pole Studio in 2003; and KT set up Vertical Dance in 2005 (but mostly they were already teaching it). Rachel candidly said,

> I am not really sure whether I invented it or whether someone phoned me up and said 'I quite fancy it for a hen party', and I thought 'yes, of course – what a brilliant thing to do before you go out on a girlie night out'. So I think it was sort of a general realisation about the potential of it.

This, from my conversations, would seem to be the most logical explanation: that, like the Victorian race for steam, people in different parts of the world were, at the same time, coming to the same realisation about pole dancing and pole classes; ideas do sometimes seem to appear from ether and occur to several people at the same time. Alison describes the early days of Pole People, one of the first pole schools:

> I was primarily just excited about this dance form, like fitness for me, in my head, came slightly afterwards – what immediately impressed me was the aesthetic, how beautiful and stunning and acrobatic it is – and I was just a bit blown away. ... I remember we had a little piece in *Metro* [a national free newspaper given to commuters[1]], it was 7th January a few months after we started, so 2004, and we had a tiny little mention but we filled up our classes in one morning, a set of four courses, all full, and we had to find more. All from one tiny mention. Whatever it was, it just struck a chord.

Bobbi and KT also mentioned how the media helped them establish their classes:

> I started as a professional dancer overseas, a trained dancer, in Japan actually. ... I did that for a few years. ... I just needed to break out and somebody said 'well why don't you try table dancing?' and I went okay, scared, but I'll try it. ... Gosh, it's not what I learned at dance school, so I went into the table dancing clubs and started doing exotic dance which was great for me because it was my own style, and then that's the first time I had seen poles [but] I wasn't interested and then when I left Japan I came back to Sydney and for some reason, took a fascination for it and started to use it a bit more in my stage shows and people started to call me Bobbi the pole dancer because no one else really used it here and then one of the clubs asked me to teach their in house girls, their other strippers in the club and I thought okay, I can show them. ... That's when it spread [end of 2003] and MTV showed me teaching some [non-stripper] girls and then came in commercial television and then it went mad, it just went crazy. (Bobbi)
>
> I went to my local newspaper and I said we're going to be starting up pole fitness classes, this isn't about ... this is totally removing because from the outset I always wanted it to be fitness based. I wasn't a particularly sexy dancer and I ... you know, so I just wanted to keep it fitness. So I had my fitness base background so I knew about warming up and stuff like that and I just said 'right we're going to, we're trying to break the mould'. (KT)
>
> I am an events organiser really. ... I did the first classes for amateurs – there was one place in London that was doing classes for lap dancers. But I thought 'if I want to do this, there are going to be other women like me who want to do it', because it is just a laugh, it is fun, it is easy exercise – it is ideal for me. ... I just got totally dropped in at the deep end. So this class – it was all new people, they didn't know I wasn't the proper teacher. I got away with it – spent all week practicing. I was just keeping a bit ahead, and then I found someone else who was really good and so I went and learned more off her. (Rachel)
>
> I think we were pretty much the first, well as far as I am concerned, we were the first people in Australia to come out with pole fitness, it is interesting because I registered pole fitness as a trademark in 2005.[2] (Jennifer)

Jennifer had a background in classical ballet, which she had trained in from age five and worked in until her mid-twenties. At the time of the interview she owned four pole studios in and around Sydney. For a while she stopped being a professional dancer and did other things which, in its turn, led her back to dancing and to pole:

> I ... worked for British Airways and that took me to London so I lived there for a couple of years. And it was actually there that I did pole dancing [classes] it had just kind of come out of the strip clubs and a girlfriend of mine wanted to do it. ... That was in 2000, it was around September 11th actually, and it [pole classes] was like at a strip club in Mayfair and it was like one pole and like pretty girls and it was part pole, part lap dancing, which didn't interest me at all, it didn't turn me off but it just wasn't my thing, but I had really enjoyed the pole. ... And it was only a 6-week course and then I came back to Sydney and did classes here – actually I did classes at Bobbi's [which] weren't here when I left.

In this way we see that pole classes grew as a result of word of mouth, the media and other pole classes; people attended one type of class and had views about how they would rather do it and so set up classes of their own with their own approach. Alison commented that 2007 saw 'the biggest boom' in the growth and popularity of pole classes: 'I did some research [in summer 2007] ... because we now have our own instructor course which we had accredited by REPS [Register of Exercise Professionals. Other instructors, for example, Elena, have also done this], which is very much for the fitness industry – and for that I counted, you know, and I think there were, then, only 59 schools in the whole of the UK'. I have not been able to complete a definitive count of how many there are now but at last count there were well over 100 schools or individuals offering pole classes across the UK and countless others in other countries.

Definitions – exercise or dancing?

The transcripts revealed two thematic layers: one was of repetition (where many of the participants spoke enthusiastically about the benefits of pole) and one was of contradiction (where many of them talked about two different approaches to pole, implying in some cases that one type was preferable to the other for a range of reasons). Within these two discourses the participants positioned themselves, and others, in relation to or against the key themes by naming themselves as 'dance' (that is, 'strippery' or high church) or 'exercise' (that is, for fitness or low church). In addition, there was also the third small category of women who just wanted 'pole bedroom', that is, to learn some basic pole moves, but primarily linking moves such as body rolls, in order to titillate their partners at home.[3] This latter group was in a definite minority but several instructors mentioned them and, while stating that, of course, it was their choice, in general they felt little respect for this approach to pole. As Rich (2005, p. 497) notes, 'people move through multiple positioning during various interactions'. Like Rich's (ibid., p. 499) participants, the polers 'were not only alluding to social change, but constructing a particular subject position of women as "free to choose"'. The participants were also adopting the subject position of poler. Their subject position shifted: they were all 'for' women and 'for' pole but, within that, some were 'against' the 'other sort' of pole and this is the key area that they placed themselves within, to or against feminism. The 'exercisers' in particular use narratives about fitness and body image which utilise feminist rhetoric, but the 'dancers' also use feminist rhetoric to discuss body image and body confidence, which is something I return to later in this chapter. Most of the instructors acknowledged the differences in approach:

> Our style, pole styles are so different, it's almost impossible to make it all the same, we can't, one person's style is so different from the next, I can't change my style to suit some wording that's going to match in, it's not about that. There are all different forms of dance. In pole dancing there is a pole but there are different forms of pole dancing too. ... There are some schools that their advertising is 'its not sexual, we don't wear stilettos, you can wear a tracksuit, it's for mums and grandmas', that's their target audience which is great, that's fine, I don't mind at all, so I am really glad we've got a bit of an edge, that I go, no, we want sexy, we want this and we want you to wear next to nothing. ... I like that we get those people. (Bobbi)

> SH: So, do you prefer pole exercise or pole fitness or what do you call it?
>
> KT: Combination of both I suppose, pole fitness and exercise, a bit of both. ... I call it, it's called vertical dance because that's what we do. And that's why we named it vertical dance ... it's totally removing it away from any pole or pole dancing in the same context as anybody.

> I think I am in the middle [between strippery and exercise], because I encourage the athletic side of it because it is brilliant for building ...

> toning your body, amazing – and through that, I have found that through the toning of the body gives the ladies empowerment and self-confidence. It is nothing to do with being able to wiggle your body around the pole. I think it is just what they can see their bodies developing into that gives them the self-confidence. And the fact then that they can say to friends who say 'you look amazing', they go 'well, I am pole dancing', and it is that that gives them the confidence. (Jess)
>
> You know what, you see the different styles too between the dancer and the gymnast and an ordinary person. Everyone has different ways of showing a sexual move and you can see the difference. ... I like the sexy bit of anything, I love watching movies that have sex in them and I love [pole classes] because no matter what age you are, you can feel sexy about yourself. (Tia)

Regarding 'high' or 'low' classes Alison explained that her classes attempt to offer a range of experiences to suit the pole student:

> We've developed a pole fitness course which is purely for instructors and there's nothing sensual or dancey about it – which kind of breaks my heart because I love dancing ... – it's pole fitness and if people want to learn just that then great, fine. But then there are people that want to keep a sensual aspect to it, and that's fine too. ... I don't want to have just a certain image. ... Some of my teachers are very fitness-y, you know like kick boxers, they don't have that much dance background – although I do like it to look nice, pointed toes. So if you just have someone, you know, that's just doing the physicality of the tricks and there's no elegance there, well, I think people need to be inspired and that wouldn't be inspiring, so we wouldn't have that sort of instructor. ... So some of our teachers are fitness-y and they really keep it like that, and others are more dancey, and that's fine, it's different styles.

The difference between styles was raised over and over again and is one of, if not the, key issue for polers. For example, Genevieve, who calls pole 'pole dancing' and wishes to keep it as pole dancing because it better reflects its creative and artistic history, explains 'if I'm talking to, you know, like people who have another instructor who does pole fitness, if I'm talking to someone who's approaching it [in] a different way, I would refer to it as pole, because then I'm not imposing my ... style or my view on them'. But Genevieve's approach is uncommon and some instructors, such as Jennifer, would like to see more distinction between strippery and exercise styles if only so that outsiders could better differentiate the two:

> [So they call it] pole fitness and yet they are still promoting themselves in a pair of hot pants and stripper heels and it is just a very mixed message. I think that that is why people are still really confused about the whole thing because the message that is being sent visually is not kind of like matching the you know the [reality], so people are still very confused, especially the media. Obviously the media are going to, you know, jump onto what is controversial and what is going to cause scandal, so whenever pole dancing is in the media it is always focused on the smutty side of it.
>
> (Jennifer)

'High' or 'low' is not the only difference between classes and schools. For example, some schools prioritise tricks over full dance routines containing linking moves. One instructor said of another, 'well, I know she prefers to do just tricks – whereas I think, it's a bit boring to watch, you know – and she doesn't smile much, I don't know why'. There are also some differences between the names of tricks/moves, for example, the 'barbed wire' is also known as the 'Jamilla'. At Bobbi's pole studios the more advanced students are encouraged to adopt performance names, to encourage them to step outside their usual 'personas' when they come to class. Bobbi actively encourages students to find or rediscover a sense of their own sensuality:

> SH: Would you say then that your classes do make people feel sexier, is that part of the philosophy?
>
> BOBBI: Oh yes, a huge part of my philosophy, I really encourage stilettos to dance in, so I really encourage, whatever you feel sexy in to wear, I don't mind what I see in there, it really doesn't bother me, I see a lot, and a lot of the moves that we do, they are very sexual as well as quite acrobatic, but I just think it's really important to link it all in together, not just make it a monkey boot camp type of thing, up down, up down and do this, so sure you can do a trick but you feel much better in yourself if you can blend it and put it to a piece of music and then it becomes a form of dance.

Other schools or instructors, as we have seen, would conversely say that they want to help their students find or rediscover their physicality, strength and sense of self. Personally, in the end, I wonder just how different those two things are – which is not to say I believe that people are enslaved to their sexuality, nor that we can easily put aside our sexual natures (thousands of years of evolution will see to that). But I mean that, overall, the instructors wish to help their students enjoy their bodies, however it is ultimately described. As Tia explained: 'it's gaining confidence and attracting people', that is, the sensual aspects of the 'high' classes are actually about gaining confidence, just as the fitness classes also are.

Finally, many pole schools, particularly in the UK, use static poles – which do not spin when doing moves. Pole-da-Cise in the UK is unusual in using spinning poles. In contrast, when I went to Sydney, I found that most schools there use spinning poles. As James said of the Mango Dance Studio where he teaches:

> We do both. Static pole is a little bit more restricted to what you can do. When you're spinning: first of all it looks more exciting, you can stay up longer; and you can use the centrifugal force to help you move from one position to the next.

Some studios, such as Jess's, use both. (Personally, I prefer the latter – it requires more strength to hold on but is more fun, as Sam also said in her interview.) These differences contribute to the development of pole; the wider pole community, in almost constant contact online, as discussed in Chapter 8, continually addresses the issues around both sorts of pole classes, adjusting, amalgamating, changing and innovating styles and teaching methods. Jennifer very much wanted to express her feeling that pole is going in two separate, albeit related, directions but that, in general, it is

> very much an evolving art form, I think that it is changing and I feel that there are people like in the exotic dance industry that still cling onto it the way that they know it. And then there is, at the same time, it is moving on, do you know what I mean? That is why I came up with the concept pole fitness initially, it was to differentiate it from pole dancing, because the word pole dancing always made you visualise ... the glass heels and the g-string. ... It is no longer only for the pole dancing that we knew it like in the strip clubs, it is definitely an evolving movement.
>
> (Jennifer)

Overall, instructors fundamentally agree, whether their approach is strippery or exercise, that women have to find the sort of class that suits them best. They also tend to agree that pole is, as Bobbi put it, not just 'up down, up down', but, in fact, artistry and athleticism, about putting moves to music, as well as having the skill to carry out the moves. They may not agree with the difference in style but they agree that different students are attracted to different studios for their own reasons. As I wrote twice, a year apart, in my field diary: *although the philosophy differs it looks the same*. Increasingly, there is also a common type of venue for classes, which I discuss next.

Where does it happen?

Ultimately, I saw, in total, 10 different venues for pole dancing – 6 in the UK, 3 in Sydney and 1 in New York. The questionnaire offered a wide-angle view of where pole classes were conducted: 69 said studio (dance studios, in gyms, health clubs or dedicated pole studios) and 60 said pubs (function rooms or private back rooms) and/or nightclubs. Other replies included a community centre (2 responses) and the instructor's home (2 responses). Also, 11 said they learnt, or were taught, at their own homes. Many of the responses said that they had started in a bar or pub and then later started to attend, or teach, at a studio. Many, mostly instructors, named a mixture of venues such as studio, home and pub/bar.

Stage 1

All the instructors except Rachel held their classes in pubs or bars, out-of-hours or in private rooms. My own classes were held in the back room of a large modern pub with curtains partially screening the classes from the drinkers in the front part of the pub. The experience was very different from attending a class in a sports centre or gym. At first we walked through the pub and downstairs to get changed in the toilets, which were in the basement, but this necessitated walking back through the pub in our 'outfits'. After two weeks we began to get changed at the back of the area in which we exercised; it immediately established an opportunity to chat to each other as we got changed and also fostered a much more collegiate atmosphere, like a communal changing room in a clothes shop or gym. During one class, the voice of the teacher was continually drowned out by the roars of the drinkers watching a football match on television in the front of the pub. In another class there had been a party in the room and some tiny plastic stars (a version of confetti) had been left on the floor and cut the feet of several women, including mine. The floor was always very dusty. Some of the students bought drinks from the bar although I saw no one buy an alcoholic drink in the classes I went to; the website does state that those who attend are welcome to do so – perhaps for Dutch courage. Two women smoked during classes, standing off to one side, twice during the two hours; the smell of smoke from the front of the pub was sometimes very strong (this stage of the data collection predates the smoking ban which came into force in England in July 2007). This venue, therefore, was in some ways a typical 'pub' experience (especially related to work such as Hey, 1988, about pubs and gendered behaviour) though it contrasts markedly with other sorts of leisure spaces, especially those used for exercise. Hubbard (2003, p. 270) points out that multiplex cinemas are 'essentially unthreatening, predictable and domesticated' as a leisure setting, which explains their popularity. But the classes were neither predictable nor domesticated.

The venue was in the city centre, in a fashionable area with lots of up-market bars and restaurants reflecting the number of legal and financial companies nearby. Several participants said the pub's location was convenient in that it was 'in town' (that is, in the city centre) and they worked nearby during the day and could come straight to the classes, or that being 'in town' rather than in some suburb meant only one bus ride from home. The classes were from 7.30 p.m. to 9.30 p.m. The first course of classes I attended with my friend was in May so it was still light when the classes finished; the second course I attended alone was in autumn so it was dark when I left. This was a consideration for several participants because of safety concerns, as Scraton and Watson (1998) also found in their study of women and leisure in the 'postmodern city', and Green and Singleton (2006) echoed nearly a decade later. Being in the city centre felt 'safer' in that it was well lit, there were always people around and it wasn't far to get to public transport or car parking. So while its position made sense to them, the fact that it was a functioning pub (that is, it wasn't closed to 'the public' on the night of the classes), with drinkers only a few feet away behind a curtain that never closed properly, was more of a problem. The curtain, and its failings, were of particular concern. Chrissy said,

> I was very aware of it at first, I kept thinking they would peep through cos, you know, well it is pole dancing and you think they will be curious.

Several others echoed an initial anxiety, for example, Grace said,

> I had one eye on [the teacher] and one eye on the curtain for the first three weeks!

I witnessed different women, at different times, march over the curtain and twitch it shut when it was offering a potential peephole. Arguably, being aware of being the object of attention is common to women, especially from their teens onwards, as Adair (1992) and others have argued. For example, James's (2000) study of Australian girls and their embarrassment at being 'looked at' at the swimming pool echoes many similar concerns. This anxiety did pass though, because as the weeks elapsed no one made any attempts to 'peep' and, in fact, showed little or no interest – it later transpired that the classes had been running in the pub almost continuously for over a year and so presumably any initial frisson of interest from pub-regulars or staff had long ago waned. Additionally, the bar manager was vehement that he wouldn't allow anyone to just wander in. However, there was some ambivalence about the pub itself, from the dusty floors ('my feet were black afterwards', said one participant) to the smell of smoke ('whatever you wear always stinks [afterwards]', said another). Leisure spaces are not just gendered but are also racialised. When we emerged from the classes I only ever saw white people, predominantly men, drinking in the pub. The very fact that the classes were held in a pub may have automatically disenfranchised different groups of women, not just because they were women, but also because of religious or cultural or personal beliefs or practices. I discuss the complexities of those particular issues further in Chapter 6.

Stage 2

Stage 2 saw a significant shift in the type of venues with more and more schools or instructors choosing to establish a permanent dedicated studio. Although I saw studios in city centres, in suburbs, in the countryside and in a country town, there were similarities between many of the studios. For example, a wall of mirrors as in any dance or fitness studio; a row of poles in front of them; pale laminate 'wood' flooring; a rack of short shorts or a shelf of stripper shoes or boots; and some extra touch of glamour or kitsch, such as feather boas in all colours piled high on a hatstand, velvet cushions, a leather sofa or a chaise lounge. One studio had lots of large vases of flowers, and Studio Verve had long burgundy velvet curtains between the foyer and the studio. My field diary describes Pole-da-Cise[4] as

> a dance studio within a hair and beauty salon, much bigger than it looks from outside. It has 5 poles, all spinning faced, by a wall of mirrors. ... It is in the middle of a housing estate next to a Spar shop [grocers], they get complaints about parking or noise. There are

> white walls, a music system playing anything from opera to R&B, black and white photos on the walls of pole students. ... It's a very different atmosphere to [my stage 1 classes], not least clean blonde laminate floor!

Nine months later, reporting my first visit to Bobbi's studio in central Sydney, my notes say:

> Everything is pink: the nets at the windows, the lights, the seats, even the keyboard at reception. A 4th birthday sign. A stage with 3(?) poles on it and more placed centrally. Mirrors at the stage and one wall. 4 windows. ... The foyer has a Barbie doll on a mini-pole, a rack of hotpants which are made to order, and rows of shoes which are sold on the 2nd floor [of the building, Bobbi's is on the 4th floor].

Two weeks later I visited Bobbi's studio in Miranda and wrote:

> Like Pole-da-Cise the studio is in a residential area so they get complaints about parking, noise, etc. ... Everything is pink as at Sydney, pinker if that is possible. It is a much bigger studio than city though, and split level. Long wall of mirrors. One of the biggest studios I have seen. I count 14 poles here, all spinning, including the ones on the stage, with much more room between each pole.

One significant similarity between many of the pole studios is that they are often in an unusual building: for example, The Flying Studio is in an old converted mill building, as is Holistica; Zebra Queen was in a shop during stage 1 of this research, later in a disused cinema and is now in Rachel's home; Bobbi's at Miranda is in a large warehouse-type building with an outside lavatory; Studio Verve is in a large business unit in the garment district of Sydney; and Pole People has just moved to a large Victorian business unit behind Regent Street in London. Parking was often an issue and something that instructors had to consider when taking over new premises. The venues in stage 2 particularly underline the anarchic element of pole classes, which I mentioned above; they give the classes a further novelty and uniqueness which says a lot for the creativity of the women who run them.

Finding joy?

This brings us to an important question: why do people choose to go to pole classes and why do they keep going? Often participants would say

they had seen it in a magazine, a friend had mentioned it or they had read a newspaper article. For example, Gidget's husband thought that pole classes might help her overcome a low period in her life:

> There was an [interview] in the local paper with Bobbi, and my husband reads absolutely everything that comes past him, and he read it and he said, 'you know I think you should give this a go because', he said, 'it's sort of like gymnastic based' and he said 'you need that'. I was working full time as a financial planner, really stressed and the kids were a bit younger and he said 'you need something that's away from all of us', because I was miserable with life.

Over and over again in the interviews women said to me that they had found a respite from their home or work by going to pole classes and that they loved pole because it reminded them of being a child; swinging on a rope over a stream or climbing monkey bars in the playground at school. There was a longing, even wistfulness, for the physical freedom of that time, and pole had recaptured some of that physical exhilaration. The unexpectedly enjoyable physicality of pole classes had restored the participants' confidence in their own abilities. Similarly, Anna Aalten (2004, p. 271) asked ballet dancers why they did ballet and one replied that

> '[t]here are rules and to me these rules always offered a challenge. In ballet I was able to test my own boundaries'. Other dancers had similar stories. [Another] spoke of the importance of ambition and the chance to excel. 'I always dreamed that I could jump over the school. It was great to work so hard with your body and to realize that you had become better. That is a great feeling, really addictive'.

These kinds of rationalisations echo many of those which I heard from the polers, for example, the words 'liberating' and/or 'fun' appear in every single interview. Unlike Rich's (2005, p. 503) participants who wished to hold on to their femininity and were concerned that their sporting prowess would make them appear butch, and unlike female body builders who share the same concerns (Mansfield and McGinn, 1993; Tate, 1999; Grogan et al., 2004), pole does not offer the danger of 'butching up'. In fact, with its residual 'sexy' image and its tendency to produce strength through lengthening and toning of muscles, pole is the ideal hegemonic, feminised sport. But add to that its primarily women-only classes, pole jams and online communities, and pole begins to look very pro-woman, if not even feminist. The relationship between instructor and student was very important and I return to it in Chapters 7 and 9. For example, most of the instructors knew why the students had begun the classes as Bobbi and Shona illustrate:

SH: Do you ever talk to them?
BOBBI: Oh yes, I know them very well.
SH: Do you know why they started doing it?
BOBBI: A couple of them started doing it because they're honest, they want to be sexy for their husbands, see that's the thing, when we first meet the beginners we get them all to say why they're here and the younger girls go, it's just fun, just fitness, just something to do and the older ladies are like, well my sex life is not so great lately, so I just thought I would like to try this, it's so funny, they're so funny. I think that's why I bond with the older students so well because they just, they've got nothing to hide, so that's primarily right, they want to be sexier.

They say they saw it in a magazine and it was just like 'oh yeah, something to do', like it's all casual, lalala, then you get down to it, by week 6 you get down to it, and you find out they feel shit about their bodies, or their relationship is in, you know, it's shit, their sex life in tatters, or their lives are empty and pole – somehow because pole has this lap dancing tint [*sic*], they think 'well, maybe it will help', they think it will change things, it's not like someone deciding to do aerobics which is pretty mindless when you get down to it, they actually start it because they actually think it will, sort of, mend, bring something back. (Shona)

I do not know the attrition rate for the pole schools that I visited so, of course, pole classes are not going to 'mend' the lives of everyone who tries it. But the people I spoke to believed it had, to whatever degree, restored something in their lives and this, to a great extent, explains why they continued to attend classes which are, in the main, more expensive than traditional exercise or dance classes. But, as I pointed out earlier in this chapter, pole classes and instructors are constantly adapting and this also explains its continued appeal:

It will work for a little while because it's a great fad if you are providing for this week but if you can't carry on, if you can't keep coming up with new material, new tricks, new routines, just changing as

> you go, if you can't do that then girls are going to go, what are we paying for? I'm doing what they did last year, you're not giving me anything new, so I think that's what I've got. I've got the creativity in the background to keep going, so girls will keep coming because they know they're going to get more each time. (Bobbi)
>
> It is a big leveller for everyone. I think it is good for those girls to realise that it is not all about what you look like. (Rachel)
>
> You don't stop learning, you keep improving, you keep making friends, it might plateau once in a while but, you know, it keeps its, er, it keeps changing. (Charlotte)

Others, such as Carole, Janice, Libby, Keisha and Ruth said they could not imagine not going to pole classes any more – that pole had contributed so much to their lives that its absence was inconceivable. Catherine M. Roach (2007, p. 31) found that the exotic dancers she interviewed began using the pole because it was, primarily, fun:

> The dancers who specialize in pole work do so, it seems, for the sheer pleasure of it, for the professional competence of mastering a skill, and for the thrill of flying. ... In any other context, the athleticism, grace and daring of these moves would qualify them for Olympic event status.

Her findings echo Aalten's quote earlier, about why ballet dancers began ballet; the feeling of wanting to 'jump over the school', and both resonate with what the polers said to me about how joyful they find pole classes:

> I am quite a structured person, you know my hair is always neat and everything about me is sort of neat and structured and I am very organised. And this is something where it is just fun and I have never done anything that you can swing around a pole. (Lilia)
>
> One of our teachers likes to say it's a climbing frame for adults. (Alison)
>
> A lot of my students were saying they feel like a kid playing in the playground and I think that's part of the attraction for them because, you know, you're supposed to behave in a certain way when you get to a certain age and I think they just feel this is their time to be free and play, you know, what you're not supposed to do anymore when you're a grown-up. (Genevieve)

> Who doesn't like swinging and climbing?! It's like being 7 years old again. (Tij)
>
> The average age group, I would say, of our members is 30+. And so the last time they did handstands and cartwheels and went upside down on the monkey bars, it was in the school playground (Evie)
>
> It's like being a child again climbing trees or a climbing frame. (2/F34/UK)
>
> So much fun, like being a kid again in some ways, the same sort of fun as swings and climbing frames – plus it gets you fit and toned very quickly. (130/F22/UK)

This type of response to pole classes is not unheard of when anyone talks about a leisure activity they enjoy. Nonetheless these joyful reports of liberation, fun and increased physical confidence begin to place pole classes on a wider canvas, whether 'high' or 'low' in approach, as something which, as Jennifer said, is evolving and meeting a need in the women who choose to attend them. The next chapter, Chapter 5, focuses on what women wear to pole classes, a point of interest in most of the interviews, in most of the conversations I had with non-polers and in most of the media coverage that I have seen.

5
What Not to Wear

P. E. kit, and how it disenfranchises or inhibits girls, has been the subject of feminist work (for example, Henderson et al., 1988; Scraton, 1992; Biscomb et al., 2000; Kirk et al., 2000; Whitehead and Biddle, 2008). Negative experiences related to physical activity at school can result in, or contribute to, a lifetime of being a non-exerciser, as many of the participants of this study testified. In 2008 Olympic gold medal winner Dame Kelly Holmes launched a campaign in the UK to change girls' school kit, arguing that more girls would continue with sport and exercise into their teens if the kit did not exacerbate issues around body image and confidence (http://www.guardian.co.uk/uk/2008/jan/27/schoolsports.schools, accessed on 2 July 2009). Whitehead and Biddle (2008, p. 253) found that

> [t]he girls did not want to work too hard and wanted to feel that any physical activity they were doing was purely for enjoyment's sake. They were generally not concerned with whether physical activity was good for them and certainly did not care for the repetitive nature of fitness-enhancing activities. For them the ideal activity simply involved having fun with their friends, and if they happened to be increasing their fitness at the same time, then that was an incidental side effect of what they were doing.

Their findings echo exactly what many of the participants of my study said, as discussed in the previous chapter, and indicate that attitudes to exercise are, indeed, ingrained at a young age. What women wear to pole classes excites more attention than just about anything else: it has excited attention from some feminist academics who believe that pole classes objectify women; from the media who have a prurient interest in what women might wear there; and, in my experience (as I discuss in Chapter 1), from a surprising number of my colleagues, primarily male, who were interested in whether women wore stripper shoes. However, within the pole community, there is mostly agreement about what should be worn. Most pole school websites offer FAQs about attending class, clothing and moisturisers. Here are three examples:

> **What do I wear to classes?**
> It is recommended that you wear workout shorts and T-shirt or vest top, as skin contact with the pole is needed. Stretchy material is best so that you have a good range of movement. If you are coming for a taster session or your first lesson you may prefer to wear workout pants over your shorts until you feel comfortable. Please bear in mind that you may be limited in what you are able to achieve without skin contact.
>
> **Why can't I apply moisturiser or tan before coming to class?**
> These are never fully absorbed into the skin, will transfer onto the pole during use and will hinder your as well as other's grip. Moisturiser is the sworn enemy of pole dancers (only while pole dancing, obviously!) and makes the pole slippery and therefore dangerous. Please be considerate and do not apply on the day of the class or alternatively shower first.
>
> **Why do I have to remove jewelery?**
> Metal against metal impairs your grip. We need to maximise your grip. You will need to remove rings, bracelets and watches. Also, the poles are specialised equipment; jewelery makes pretty patterns on the poles but also damages them (the poles that is; though your jewelery may also become damaged).
>
> **What about shoes?**
> Footwear can be split sole jazz shoes, ballet shoes, trainers or similar. This is something that is personal; you will soon find what suits you. You may have the opportunity to wear heels later on in the course.
> (from www.flyingstudio.co.uk, accessed on 1 June 2009)
>
> With a Vertical Dance class the pole requires skin contact to cause traction or a break, if you do not have bare skin traction to the pole can not be had and the body part can not grip causing too much slide, so we recommend sports shorts that are tight fitting around

> the upper thigh area (cycling shorts are too long) and a vest top, how ever if you feel uncomfortable please wear loose fitting tracksuit bottoms that can be rolled up. Please do not wear any body creams or lotions on the day of your class. Cream coats the skin in grease which transfers to the pole making it slippery and dangerous.
> (from www.verticaldance.com, accessed on 1 June 2009)

> The most important thing is to wear clothing that is non restrictive and that you feel comfortable in. However, you'll benefit most out of your class with bare arms and legs. We would recommend a pair of shorts (better to grip the pole with!), singlet and runners. We ask that you do not wear any oils or creams prior to your class as this will impede your grip on the pole.
> (from www.studioverve.com.au/faqs.html, accessed on 1 June 2009)

Shorts, jogging pants and T-shirts are not a departure in terms of an outfit for exercising. These examples indicate that the instructors take both safety and the comfort of their students seriously, and that classes are not as salacious as some would like to believe – yes, particular clothes are recommended, but just as they are for any gym or dance-class attendance. Some bare skin is required to adhere to the pole and moisturiser is a no-no because it inhibits adherence to the pole. But apart from that, as Rachel and the examples above stress, women are encouraged to wear what they feel comfortable in:

> You have to have bare legs – but other than that ... I think you should wear something that makes you feel good. ... It makes it a lot safer because people get a lot better leg grip, so it means people can get upside down quicker and they have got a lot less chance of falling off once they are there.
>
> (Rachel)

Some perceptions about how students might dress for a strippery class can be misleading as Bobbi and Sara pointed out:

> We got sort of labelled as the Barbie Doll Pole School. ... We heard once that someone said if you go to Bobbi's you have to have full make up on and your hair done. We were like 'what? What?!' (Bobbi)
>
> It's not like everyone is in, you know, sequins and feather boas all the time, sometimes it's just sort of normal stuff, little gym clothes, like you would get anywhere. It's only for the [end of term] performance

that they get dressed up and even then it's their choice, some don't, some do. (Sara)

JESS: I have got the feather boas and the shoes, so for the ladies. ... I mean you can spot the difference between each lady – some people want to come just to strut round in high heels and shorts, where some people come in, self-conscious, don't want to wear shorts – 'if I am having to wear shorts, I am not coming', sort of vibe.

SH: And what do you do about that?

JESS: Again, it is fine if you want to wear [long] pants – I do it in jogging pants and roll them up and I say that is absolutely fine. As long as you have got your shins bare to grip on the pole; that is all that matters. But if you want to come in shorts, I have got the shoes – you can wear the shoes and you can strut around and have a play, you know.

We can see that pole classes offer a choice of what to wear, which many participants, particularly the non-exercisers, had not previously encountered in their negative experiences of physical education at school. 'Doing what's right for you' is, again, important and although there is the greater likelihood that you will be offered, or advised, to wear stripper shoes at a 'high' class, if you are there, presumably you have taken the decision to attend that type of class in the first place. Tia chose Bobbi's in just that way:

TIA: I've been doing this for 4 years, this will be my 5th year into it. ... Back when they first opened up Bobbi's Pole Studio ... I got on that phone so fast! I thought they're opening up a club and I will be the first one there.

SH: So were you excited, or nervous or?

TIA: Well, when you went there, they said 'look we're going to show you what they do' and when I got up there I thought 'oh god' because there were people with their high shoes and they had their little skimpy things, and I thought, 'what have I got myself into?' ... I have improved but, yes I'm wearing skimpier little things, I don't like to put my pants right up my bottom. ... But even the other students who are big, they tend to wear pants that are right up their bottom. Like, really big girls and it's just unusual to see. I don't mind flashing my tits though, I do it all

> the time, my family are flashers of boobs, we like low-cut tops … and I don't mind using my arms because I've toned them, you know. I'm not a fat person but I still wouldn't show my belly.

The progression to 'skimpier' clothes for class, which we will examine next, was mentioned in most of the interviews.

From baggy shorts to hot pants

Frew and McGillivray (2005) argue that 'physical capital' in gyms and health clubs keeps the exercisers in constant anxiety about their shape and size. Tara Brabazon (2006, p. 67) describes gym culture as a 'site of both anxiety and display'. Similarly, Brena R. Price and Terry F. Pettijohn (2006, p. 997) note that 'working out in tight clothes, especially at a public health club with its wall-to-wall mirrors, may hamper performance and threaten body image'. Pole classes contrast these with their findings that the participants became more confident about their body shape and size, more willing to wear shorter shorts or cropped tops, as the weeks went by. As Price and Pettijohn (2006, p. 994) found, attire does have an influence on how people feel when they are exercising. In their study of female ballet dancers, the dancers 'reported significantly more positive body and self-perceptions' when they wore 'junk' such as sweat pants or baggy T-shirts than when they wore leotards and tights. They also felt, and performed, better when there were no mirrors. There are some echoes and some contrasts with the participants of the pole classes that I attended during stage 1 of the research. For example, in the pub, there were no mirrors, only other women to reflect one's performance, and the atmosphere was one of unremitting support. So although most of the women began in what Price and Pettijohn described as 'junk', they quickly started trying out different sorts of outfits that they perhaps had not had the confidence to wear previously. Ruth's explanation was typical of the students:

RUTH: I realised that not everyone had a, you know, like a really fit body, we were all sizes in there but still we all looked good when we were on the pole so it gave me more, um, more confidence to wear stuff that was more, er …

SH: more revealing?

RUTH: Yes, I bought some hot pants and I felt OK in them. I think we all mucked in together, if you know what I mean, we were in it together, we all had damn cellulite, but our confidence grew as we got used to each other and as we got better on the pole.

It's so strenuous, you get so hot, so of course, you would go the first week and afterwards think I'm not wearing a T-shirt with sleeves [again] because I just sweat profusely, I'm going to wear a singlet, so of course the second week it was a singlet and then you know, you felt, oh those shorts are too long, and they're getting in the way of doing things and you start pulling them up and then I thought I've got shorter ones, I'll just wear shorter ones and then week 8 was, oh I had stupid heels. And then I thought, well I have just a strappy singlet top, I still hadn't exposed the belly but I just had very, like underpants really with lace on them for week 8, so yes, it all came together. (Gidget)

The common reaction was that, although few of the students had a 'perfect' slim, toned body, it ultimately didn't matter. Body shapes and sizes were accommodated and overlooked and rapidly became irrelevant. This contrasts sharply with the work of Wesely (2003) whose participants were professional exotic dancers who felt obliged to seek the 'Barbie' look they imagined their clients demanded, and to achieve it they used a range of body technologies, from taking drugs in order to lose weight, hair dye, blue contact lenses, breast augmentation and other surgeries. In the classes I attended the initial comments such as 'oh I'm too big [or unfit, or old] for this' or 'how can I get my fat arse up there?' which I had noted in my field diary in the first few weeks, vanished altogether. Being a 'Barbie' may be required for professional pole dancers but it wasn't at all required for pole classes. Most of the instructors discussed the changes in women and their clothes as the classes progressed, for example:

Oh yes, their heels get taller, their pants get shorter, it's such a big deal for women to come in and put on a pair of hot pants and that usually takes a couple of weeks, particularly for older women, you know, they are middle-aged, they have a couple of kids, their body confidence has taken a knock. ... You are not only like improving their body, it is not even so much about that but improving their strength and their body shape, improving their body image and their self esteem, and that is part of my job and I love it. (Suzie Q)

I always say to the women at the beginning of the class – 'what do you do'? [and they say] 'I am just a mum' – I am like 'you have given birth to children, you look after their lives, you look after your husband or your partner, you organise the house ... there is so much that you do'. And at the beginning of the courses they come in with their shorts down to their knees and massive big baggy T-shirts on. As the weeks went by, they would be turning up and the shorts are getting shorter, and they are like 'yes, I have got something', and I say 'of course you have – it's not just because you are on a pole', that is very much a side that they let themselves go. (Sam)

SH: So do they often start in one sort of outfit and another by the end of the course?

BOBBI: Yes, smaller and shorter, it's fine, I don't mind. Most girls will start with like little bit longer shorts and then as the terms go on, the shorts get shorter and shorter, but they do that because it's more practical and they get more confident with their body as it starts to look better, so they show it off a bit more, so it's up to them. I encourage it because it's easier on the pole, it's hard for me to instruct a move when a girl's got shorts down to there when she needs this piece of skin, so I just say, without being, I don't mean anything sexual by it, I'm actually pleased you've got to hike them up, you've got to roll them up and if you just say it like that, 'hike them up!' they go okay, don't even think twice and up they go because otherwise they can't get on the pole and then one day they go well I might as well wear short ones anyway because I'm going to be rolling them up and the next thing you know they've got short ones on.

Clearly, at the point of being a beginner, the issue in Bobbi's explanation is not about her personal preference for them to wear short shorts but about the student being able to accomplish increasingly skilled moves, in order to progress both in their abilities and in their self-confidence. But the spectre of having to wear stripper shoes or tiny clothes may put women off attending, and this is how more exercise-focused classes came into being, as Jennifer explained:

I thought there was a hole in the market for women that would probably would like to give it a go but would be totally intimidated by the exotic dance thing. ... I thought 'surely if I feel this way there must be other women that feel this way', because I did not want to be in an environment where I felt that I had to take my clothes off, which is what some of the other schools are; even though it is not verbally said it is kind of the energy and the atmosphere and then you almost felt that in a very subconscious way manipulated into doing it. So you kind of desensitize because everybody is walking around in 6 inch glass slippers and hot pants and I just didn't feel comfortable with that and I thought 'well if I feel this way there must be, you know, others who do too'.

Footwear is, as we saw in the FAQs at the start of this chapter, an ongoing issue and very much contributes to the image of pole classes, both among the pole community, and more widely.

'High church' footwear

Footwear is one of the ways that pole exercise classes are distinguished from other sorts of exercise classes in the popular imagination (such as those worn on the cover of this book). But any item associated with women and sport or activity can become both iconic and sexualised, as Jaime Schultz (2004) argues about the sports bra. Catherine M. Roach (2007, p. 33) describes the first time she tried her friend's stripper shoes on:

> I take my first tottering steps. The shoe has a stiletto heel, six inches high, with a steel rod built into the heel to prevent it from suddenly snapping – every dancer's fear. ... The circumference of the bottom of the stiletto where it touches the floor is considerably smaller than that of a dime. The front part of the shoe is two inches high so that the rise between the ball of the foot and the heel is four inches. The shoes turn many girls [*sic*] into six-footers, or more. ... You can't go heel-toe as you normally would ... you have to put the toe and heel down at the same time. To make this happen, Marie demonstrates how you have to tighten your bum muscles, tilt your pelvis, and sway your hips.

These are the same style of shoes used in many schools, mainly strippery schools and which, a few years ago, were considered the norm for pole classes – but that is because they were considered the norm for lap-dancing clubs. In the past three or four years the idea that heels, of 6″ or lower, or bare feet, jazz shoes or trainers (sneakers), are all suitable and acceptable has become the norm instead. For example, in my interview with Alison, on looking at the cover of the *Pole People* instructional DVD, I commented that all the women on the cover were wearing stripper shoes and Alison said 'oh yes, but if I did it now they probably wouldn't be'. As I noted in the previous chapter, pole continues to evolve. Footwear has been a key part of that evolution.

In stage 1 of the research, only one teacher (Rachel) refuted the idea that high heels were necessary; she called it 'tottering around'. In the classes that I attended, most of the women bought 'fetish' shoes off the Internet or from the company that runs the classes. These shoes would have straps, heels of at least 6" (15cm) in height, and possibly also a wedge, making them even higher. I bought a red patent pair and, as someone who never wears heels, found them very unsteady and painful at first although several women wore heels regularly so didn't find them quite as difficult. Our teacher walked in hers with no problems and also wore hot pants and a tiny vest, as required by her employer. But the shoes are not necessarily a source of grace as I encountered several times when observing classes, even advanced classes, and this field diary entry from an observation at a pole studio is typical:

> All wore 6" heels but didn't walk in them particularly gracefully, lots of clumping about, very very loud on the wooden floor. Didn't look any more balanced in them than as we had done as beginners! The shoes occupied much of our attention, whether they hurt, whether we were 'stomping about in them' as Kate put it, whether we would twist our ankles in them, particularly when landing, or whether we were getting used to them. I feared hearing the wet snap of my ankle breaking every time I landed for the first two weeks which I mentioned to Bobbi who was one of the biggest advocates among the instructors for heels:
>
> SH: I was worried my ankle will snap.
>
> BOBBI: I still think that about mine still [laughter]. It's something funny about it, I can't say they help the dancing, I really don't know how, they help you hold yourself up better.
>
> SH: You're much higher up as well.
>
> BOBBI: Yes, yes they help with your posture, you look beautiful but as far as practicalities go, like I can't do anything without them on, I just can't pole dance if I'm not wearing them.
>
> SH: Is that because you learnt in them do you think?
>
> BOBBI: Because I refused to do anything looking in the mirror without my shoes on, I don't want to look in the mirror, I refuse to look in the mirror without stilettos on, once I'm in there [a class or a club], I have to be dressed appropriately, otherwise I don't see the point of doing it, I just love the whole thing.

Plasters (Band-Aid) were shared at some point during each course of classes that I attended. The teacher assured us that there is a 'point' to the shoes in that you can hook the heel onto the pole for certain moves and the Pole Stars website asserted that the heels are necessary to give students the 'necessary strut'. Various rationalisations were posited during the classes I attended and the interviews I conducted: that the heel was to hook onto the pole; that it helped you start off at a higher point; that it helped you to bounce around the pole when doing linking moves; that they elongated the look of the legs; and just that they looked special, theatrical. The classes were themselves expensive but the women were not unwilling to spend even more money to buy the shoes, but overall the attitude seemed to be one of detached interest because these shoes, novel as they were, were not about to become an integral part of these women's lives; the shoes were little more than *objet d'art*. They viewed their shoes as a curiosity, knowing that such footwear was closely aligned to pole dancing in strip clubs and so would cause some interest for that reason alone; the shoes were compared, derided and admired. However, they did also start to develop some agency about them after a few weeks: they would put them on when the teacher said it was necessary or when they felt like it, but then take them off when they felt like it, or when they felt unbalanced or when the straps were cutting into their feet. There was much laughing and semi-fake grumbling about them, which fitted in the general mood of camaraderie, discussed further in Chapter 7, but most of the students said they liked their shoes; that they found them painful but beautiful. They also dismissed the notion of the shoes being somehow 'threatening', either as an item to wear from choice or as a tool of oppression, because they were simply so outrageous and impractical that it would be hard to take them seriously. As Carole told me 'if male friends say how fascinating or sexy they find them I say "right good, well let's see you put them on and walk to [the supermarket] and back, then tell me how fascinating they are!"' Carole's brisk response was typical of how the students rationalised pole exercise's links to the sex industry by using humour or irony, and again this is an issue I return to in Chapter 7. Catherine M. Roach (2007, p. 34) found a similar attitude to shoes in her research on exotic dancers:

> While I had expected more ambivalence and resentment towards the heels, given how uncomfortable they seemed to me to wear all night long, most of the dancers with whom I talk are surprisingly positive about their shoes. ... The appeal mystifies me: a woman in very high stilettos looks to me hobbled, not sexy. But the dancers seem to get used to the shoes quickly, learn how to walk in them, and then embrace them as an integral and valuable part of being the stripper.

The difference between classes where heels were worn and classes where they weren't was mentioned by Suzie Q:

> I have to work in those fucking shoes, I like a rest, they really hurt ... but I like that strippery side of it. ... At iPole I just teach in bare feet but I think that is the difference between Bobbi's and iPole, that kind of thing. I think the big shoes are exciting and a definite part of it: it is your pole shoe, your pole outfit, and I think it makes it more of a theme, a separate world. ... I have got the real life experience and I know what it is like, it is pole dancing, and putting on shoes and pants, the moves they do, it is emulating strippers, it is, sorry. I know that pole dancing it often trying to segregate itself from stripping.

Rightly, Suzie Q points out that some women choose classes in which heels are required and some women – as Shona said in the previous chapter – choose strippery classes because they appeal to them for whatever reason, whatever life circumstance, they are currently experiencing. To me, this underlines how pole classes occupy a space which is both in contrast to pole dancing in clubs and at the same time reflects pole dancing in clubs. As Suzie Q says, what the students do in class echoes what exotic dancers do in clubs, to a certain extent, in the outfits and the moves if not in the audience or the intent. Catherine M. Roach (2007, p. 36) eloquently explains the appeal of the shoes for exotic dancers, which in some ways reflects the appeal in classes:

> It's when the topic of height comes up that I finally start to get it. The dancers talk about the shoes as sexy because they arc your body into a sleek, elongated curve and give you a hip-rocking prance of a walk, but they talk about the shoes as conferring power because they make you tall: as tall or taller than the men.

But this shift towards sartorial choice in classes shows us that classes are indeed evolving separately to pole dancing in clubs, not least because classes offer choice. The Pole People website says: 'flat shoes are fine if you're not a heels kind of lady!' (www.polepeople.co.uk, accessed on 1 June 2009) and the Pole-da-Cise website states: 'this is a fitness class so heels are not required' (www.poledacise.com, accessed on 1 June 2009).

'Low church' footwear

Not all of the instructors or students who chose to do pole in bare feet or flat shoes came from a sporty or non-exercising background, as you might expect; some had a dance background, such as Genevieve and Jennifer:

> I'd dance in dance shoes and not the big heels but then I'm tall, you see, a lot of dancers in the clubs are quite small, if they're like 5', you know, below average height, they need to wear these shoes to make themselves look taller, to elongate their legs, that's how that started. But I don't need to wear them because my legs are long already so there was me with my flat shoes, my jazz shoes.
>
> (Genevieve)

Genevieve echoes what Bobbi says, above, which is that she is so used to doing pole in the shoes she prefers that she couldn't do pole without them now, a point which Jennifer also made:

> There are two ways that you can look at this: there is this argument where pole dancing as we know it in the modern day world or western world was performed in the strip clubs and the strippers wear those heels; that is what they are used to moving around in. So whatever they are going to teach is what comes naturally to them. ... But why would I put women in 6" heels who have never been in those heels before, I mean first of all they are going to look, and feel, clumsy, and it is dangerous as far as I am concerned. Because you are more focused about being able to balance on high heels than you are actually trying to learn and do movement, and then you get caught up in the image of what you look like as well you know, so coming from a dance background I just feel that you can't work through your feet properly. When you are in a pair of heels, as everybody knows, even women that wear heels everyday end up with really tight calves and short Achilles ankles, tendons.

But mostly, if an instructor had a sporty background she would prefer not to use heels, as KT explained:

> The other thing is, a lot of the stuff that you do, I am uncomfortable doing that kind of thing – that is why I don't teach it, because I feel silly. And the majority of women are really uncomfortable about doing that. ... I believe that you should wear what you are comfortable in wearing, and that is why people will say 'no, you have got to wear shorts and a vest'. You shouldn't have to wear heels – you can't

> do fitness in 6" heels, I am sorry, no matter how much you try, you just can't. It is not safe, it is not effective.
>
> (KT)

Jennifer believed that not only did heels impact on how well a woman could physically exercise but also on the effect it could have on her confidence:

> SH: Do you, do they talk about feeling sexy?
>
> JENNIFER: Not so much because it is not really our focus here so that is the difference between what we do here and probably some of the, we don't focus on the sex we focus on the fitness. ... However, not most women but some women, that is their main focus is you know seeing themselves in the mirror and trying to match up some image they have got in their head about the way they should look. ... I guess it depends on the environment they are in, I mean if they are doing a class in a pair of heels obviously sexy is never going to leave their minds because they are doing it in a pair of heels, rather than if you are doing it in a pair of sneakers. I am of the belief that dressing sexy is like putting a band aid on a deeper problem. I think it has got to come from within, I think that if you are, say, overweight you feel bad, you are not going to feel great about yourself, if you start to look after yourself and you feel fitter, stronger, you know, you are going to feel good.

Again, we see that Jennifer's approach is similar to that of Genevieve and Alison when they asserted that they refuse to talk about sexiness in class because of how it might impact negatively on a student – I return to issues about confidence in the final chapter. Of course refusing to do so also distances pole classes from pole dancing in a club, thus maintaining an overall philosophy and approach; it preserves a certain utilitarian approach to the classes, which ensures that the message is not mixed. But we can see why the pole exercisers want to be clear about what footwear to wear: it is because the shoes certainly dominate the popular image of pole classes, as Catherine M. Roach (2007, p. 33) writes about the weight of cultural expectation which surrounds it:

> I hear many shoes stories in my interviews. The stripper shoe is such an iconic symbol of the exotic dancer that it becomes loaded with all she represents: the oppression of the industry, with echoes of

> foot-binding ... the element of fantasy, play and glamorous theatricality; a perceived low-class, vulgar aesthetic; and the possibility of get-even-with-the-boys empowerment. Indeed, the shoes' platform risers and stiletto heels embody many of the paradoxes of the profession itself: sexism and sexiness, power and vulnerability, glamour and tackiness, pleasure and pain, danger and desire.

As Roach notes, being as tall as your clients can be a plus for exotic dancers; and the shoe's hints of bondage amplify its sexualised image, which is also of benefit if you are working in a club. It is important to stress that pole classes offer choice; are not subject to the male gaze; do not require that the shoes be worn all night, if at all; and even if women do choose to wear them they often do so because the shoe is 'exciting ... [part of] a separate world' as Suzie Q said. Indeed, the women who choose to go to strippery classes are often excited about buying the shoes, as if the class has given her the opportunity to buy something she had previously felt curious about but not justified in buying. Perhaps, just as pole offers a chance to feel liberated physically, it also offers a place where you can 'play'; and as Jess said, you can dress up. I was surprised that many of the participants had an affection for, and had, as children, played a fairy-princess figure, apparently the epitome of feminine docility and 'girliness', as pointed out in *Alternative Femininities* (Holland, 2004, pp. 53–4). Yet I found that they described their fairy-princess play as 'physicality and pleasure rather than physical repression and weakness', and there are some parallels between that 'shameful hankering' (ibid., p. 55), as one alternative woman put it, and the appeal of stripper shoes, feather boas and sequined shorts for the 'high church' participants. Pole offered a place not only to connect or reconnect with their physicality but also a place to dress up in a way they may not be able to outside class. In tandem with that, having bare feet, or wearing trainers or jazz shoes, gives choice; it equally appeals to particular women while also offering a safe, women-only space for physicality and fun. Both options perfectly reflect the evolving, current pole class and indicate that pole's appeal is not as crass as some feminist or media writers would like to believe; its appeal is more complex. In the next chapter I examine some of these complexities: race, ethnicity and age in the classes; and I return to the question of empowerment.

6
Diversity and Empowerment?

> *The diversity of students in our courses encourages us to empower each other through the art of pole dancing.*
> (*www.poledancingschool.com/*,
> accessed on 10 June 2009)

This chapter focuses specifically on the diversity, or not, of pole students and instructors, looking first at issues of race and ethnicity, and age, and going on to discuss ideas about empowerment. Is there diversity in pole classes? In Chapter 4 I discussed the 'whiteness' of the venue of the classes (a city centre pub) during stage 1 of the research; I also discussed how by stage 2 of the research many schools or instructors were setting up dedicated, women-only pole studios. So while initially, and to some extent still, many classes took place in venues which may have been unwelcoming to some groups of women, overall the shifts in the venues of pole studios has made classes much more inclusive, at least in theory.

Race and ethnicities at pole class

There were 117 responses to the question: how would you describe your ethnic group? 18 did not answer that question. Of the 117 there was 1 'white and African Caribbean' respondent, 1 'African American', 1 'black', 1 'Chinese', 1 'Mestizo' and 1 'Moari'. The remaining 111 were all 'white' (3 white men and 108 white women) with some variations such as 'white/welsh' or 'white atheist'.

Stage 1 of the research had 15 interviewees of which 11 were 'white', 2 were 'black British' and 2 were 'South Asian'. All chose their own pseudonyms except Rachel who wanted me to use her own name.

Stage 2 of the research had a further 22 participants of which 3 were 'African American', 1 was 'Korean' and 18 were 'white' (17 women, 1 man). I accept Aspinall's (2002) suggestion to use terminology, which is defined and accepted by those being described. Those with pseudonyms all chose Westernised names. The fact that the participants were predominantly white is feasibly due to two main reasons: that polers themselves are mostly white, although less so in the US; and that snowball sampling was sometimes used which does tend to replicate the sample.

Methodological issues arise from a white, British researcher interviewing women from other racial or ethnic groups (Watson and Scraton, 2001). As Afshar et al., (2002, p. 9) has argued, there is still little work about the interviewer–interviewee impact on the research, 'particularly when researching across cultures'. Jane said in passing that she noticed that she was the only black woman at the classes (apart from her sister) but wasn't interested in engaging in further discussion about how 'race' was represented at the classes; she said, during our interview, 'where I go, it's always lots of white people, or none at all, so either way, it's just how it is, you know'. I felt unwilling to ask her why she thought pole didn't seem to appeal to many black women, because I didn't want to appear to be asking her to 'speak for her people' as it were. But I had encountered this particular methodological issue before: in previous interviews (Scraton, Caudwell and Holland, 2005) I had conducted with black and Asian women footballers, I encountered the same reticence to talk about race or ethnicities. In contrast, in stage 1, Kosa was particularly aware of issues about race and ethnicity, and wanted to talk about them, pointing out that the classes she attended mainly comprised white women.

> I knew it would be like that because how many Pakistani girls, or even Indian girls, although there is one, are going to go to a pole dancing class? Not many. Not just cos their families wouldn't want them to, although that would be a big factor, but also cos, you know, they would think it wouldn't be something they should do. We hold ourselves back a lot.

When I asked her why she attended, she said it was because

> I am, like, Westernised, even though my family don't like it, I mean, I will wear the salwar kameez in the house but when I am out I wear jeans and like little tops. I used to get changed when I was out but I don't bother any more – they know I go out like that, well yes, they see me, so I get a lot of stick for it.

As many researchers have pointed out, the term 'South Asian' is used in the UK to denote a wide variety of people with differences in religion, country of origin, age, sex and class (Brah, 1996; Anthias, 2002; Andall, 2003). However, Kosa here provides a very specific background for herself and her family: she said she described herself both as South Asian or as British Asian depending on who she was talking to, that she didn't describe herself as a Muslim to anyone and also that she belonged to the working class and was of Pakistani background. Kosa did encounter difficulties because she was from a Muslim family which disapproved of her lifestyle. She had distanced herself from her family in some senses in that she identified herself as 'Westernised' and didn't want to accept an arranged marriage, and also because she was often out in the evening attending various classes. She still lived with her brother and his family despite mentioning that she could afford to set up house for herself, because living with her brother meant that she had more money to pay for her leisure activities. She saw this as a necessary balancing act which allowed her an active life of leisure outside the home; if she had had a place of her own she would have had to sit in on her own much more often and would have much more spare time on her hands. Kosa was unusual among the polers I spoke to in that she saw herself as both 'integrated' into the class, and the white Western society in general, but would also mention ways that she saw herself as different and an outsider, owing to the trouble she had at home regarding her independence. Kosa was attracted to the classes because she enjoyed fitness classes, saw that they were all-women and thought it looked like 'good fun'. She often felt that the tension at home could stop her from attending, although it never did.

Tij described herself as British Asian or as British Indian. She and Kosa did not attend class together although each mentioned that they said hello to each other: 'there is another Asian girl but she is Indian, she's not like me. We say hello, nothing else' (Kosa) and, with noticeably more coolness, 'there is a Pakistani in the class, I never got her name, but we always said hello' (Tij). They did not pair up during class. Tij said very little about race or ethnicity in her interview except when I asked her if she had always regularly exercised; she said 'no', although

> that isn't cos I'm from an Indian family, we are, we are all totally, you know, we are different yes in that there are different considerations, about caste and that, and beliefs, we are Sikhs – although I am not what you would call practising – but yes, if I am in a classful of white people I don't feel funny. It is different for Muslims I suppose,

> they are much more, like, separate I think, you know, more, er [long pause] insular.

The diversity of pole classes reflects wider trends regarding physical activity and ethnic minority women. For example, Hasina Zaman notes that there are three distinct models of participation in physical education for Asian girls which are assimilation, integration and separation (1997, p. 51). Arguably, Kosa and Tij fit into at least two of these categories, in different ways. However, Zaman also argues that the structures (referring specifically to Muslim girls' experiences of P. E.) are not only patriarchal but racist (ibid., p. 53) in that the conditions, such as showers or kit, are not conducive to Muslim girls being able to comfortably take part. As I discussed in the previous chapter, P. E. kit is not a concern only to South Asian girls or, in this case, women. The issues around what is worn to pole classes are relevant to women whatever their ethnic background, just as research such as Whitehead and Biddle's (2008) illustrate that white girls also find P. E. kit inappropriate or embarrassing.

The white participants did not mention race at all, which perhaps isn't surprising as 'whiteness' remains the norm and hence invisible (hooks, 1983; Hill Collins, 1998; Knowles, 2003). I asked most of the instructors if their classes were mostly full of white women and in general the answer was 'yes' with some saying that sometimes one or two Asian or black women would attend, for example:

> In a lot of the other classes there is always a lot of Asian girls, Indian girls. I have seen a couple of Indian girls when there has literally only been 8 people in the class anyway.
>
> (Silke)

In general, the classes that I observed were predominantly or entirely white but there would often be at least one black or South Asian woman in the class, either alone (or, in Jane's case, initially with her sister Grace) or with a friend, mostly a white friend.

During stage 2, the most mixed classes were in the US. I interviewed 3 African American women and 1 Korean woman but, since none of them initiated conversation about ethnicity, I chose not to prompt them, again wishing to avoid asking them to somehow represent all African American women or all Korean women. Now that I am writing this chapter, of course I wish I had but, since the research was never meant to focus on race and ethnicities in class, I allowed the participant to lead the discussion (as I explained in Chapter 1).

Age at pole classes

The questionnaire respondents were aged between 18 and 54. The interviewees were between the age of 18 and 55 and, as I pointed out in Chapter 1, polers cannot be said to be anything if not varied:

> There's so many different types of people that we've had in our classes over the years, you know we've had undertakers, doctors, nurses, you know ... high flying people. ... We've got, the oldest person we probably had was about 62 – and I personally, this is a difficult one and age, the age thing is a really, is a tender subject and I don't really ... but just to keep everybody happy, I say 18, not necessarily that I believe that it should be 18 but at the end of the day I don't want to give them, the mouthy people, the opportunity to be bad.
>
> (KT)

Similarly, Bobbi discussed the youngest age she felt able to accept and saw it as a potentially delicate issue:

> I mean we tap it at, I mean bottom tap is 16, because we have 14 and 15 year olds who want to come. At 16 we say they can come with their parents' written consent, both parents. Over 18 obviously anyone can come and no age limit, so we get 60 year old ladies and stuff.
>
> (Bobbi)

A pole school in New Zealand runs courses for children, which I return to later in this chapter:

> At Kiwi Pole we have taken away the sleazy side of pole dancing to bring you a new and fun sport anyone at any age can get involved in. For young and old, big or small any occasion, we can teach anyone to climb and spin around a pole.
>
> (www.kiwipolefitness.co.nz/, accessed on 9 July 2009)

Kiwi Pole runs a range of classes, the most innovative range that I have seen so far:

> **HerPole** — Ladies only sorry guys. This is pole dancing for fitness, learning pole tricks and different combinations. We do mixed level (M) classes as well as, beginner (B), intermediate (I) & advanced (A) levels.

> **HisPole** — These classes are for Men only, sorry ladies. All about building strength, concentrating more on Chinese pole style of pole fitness. No dancing involved!!! Any age shape and size welcome! Mixed levels.
>
> **Couples** — These classes are for couples to come along and share a pole, help each other in training and encourage your partner. Mixed levels men and women welcome.
>
> **KidsPole** — These are for kids only! Pole play classes. **Parents must accompany** their children. Bring the kids and let them learn to climb and spin. Loads of fun and great fitness and strength building for the kids.
>
> **MumPole** — These have been put in place for us mums that have children at school or kindergarten during the day and only have that time to be able to come and do it. For any shape and size!
>
> NEW!!
>
> **Infant & Mum Pole** (I&M Pole) — A pole session put in place especially for new mums. Come meet other mums and bring your little bundle with you. A class designed to be a social fitness time for new mums and their new babies. No children that can crawl or walk around the studio allowed for safety reasons. Up to 12 mums so bookings are essential!
>
> **35+ Pole** — We have decided that because a lot of mature women like doing pole fitness we would like to give the more mature women there own place so as they don't feel uncomfortable with the younger ladies that come. Any age over 35, any shape or size, Women only!
>
> (www.kiwipolefitness.co.nz/, accessed on 9 July 2009)

They also offer open studio, pole parties and private lessons (all of which most schools do). As we see, Kiwi Pole runs classes for women with small children, fulfilling a social need as well as a need for exercise; for couples to attend together; and for children. They also run classes specifically for women over 35 so that women feel more comfortable – and confident – about attending. But most instructors had a mix of pole students:

> An amazing mixture, that is what was so good about it. And the ladies who ... people who have shops that have my [publicity] fliers,

> like shoe shops and things like that – I couldn't have believed the wide range of people who take your fliers. And I have had people from 18 to – it is fashionable at the moment to bring your granny on your hen night, you know, or your Mum – so I have 18 to 65 year olds here some afternoons. (Rachel)

SH: So what sort of age range of the women have you got?

JESS: The youngest girl I teach is 18 and she is wanting to go into it professionally.

SH: So that is why she is training?

JESS: That is why she is training, yes. And the eldest is 40.

SH: And do you know what they do?

JESS: Job, career wise? Oh yes – and it's a good thing ... but I suppose it is a bad thing as well – I am very personal with people who I am teaching so we end up being friends.

SH: I suppose that is a benefit of a small class really.

JESS: Yes, I like it – it is nice, and they like it – they obviously like it because they keep coming back. Yes, so I do get to know their careers and they get to know ... I have got some doctors, I have got some fitness professionals, so people who teach aerobics, some people are students.

Both Genevieve and Tia talk about how pole classes have provided a sense of oneself which age, in a sexualised culture, takes away from women:

> I think this is addictive and like anything some things can be more addictive than others and that's how I find it, I find it very addictive. ... I think its because its improved myself as a sexual being and to me that's how I look at it. ... I never felt like that, for years I haven't felt like that. As soon as you hit a certain age, you don't feel like that. So when something makes you feel so good, even lap classes, I took up lap classes and I felt, well, I just felt better about myself, I felt like a sexual being again. (Tia)
>
> I would guess the top level [age] would be 50ish and then a few older here and there. ... A lot of the older women like that, it's about, you know, they always tell me it's issues about reaching a certain birthday, like 40 or 50, and if I'm not going to do it now I'm never going to do it, and maybe they have the menopause, or have had breast surgery, and they want to get something back. (Genevieve)

Clearly this is a complex issue; paradoxical, in fac
patriarchal society women are made to feel increasin
age; this we know, cannot be refuted. Even the new
celebrities do not really reflect the reality of agei
do, to surgical and non-surgical cosmetic procedures. As J
(2008, p. 117) argues,

> we can use dieting, make-up and even Botox to assert our sexual selves, not necessarily just to comfort out neuroses but also as a liberating, self-conscious tool, enhancing our sensuality. There is a blurred line between empowerment and neurosis, pleasure and collusion, coercion and wilful choice. Part of our empowerment should also come from knowing how precariously we balance on this fine line.

Yet, according to the participants, attending a pole class (which has its roots in lap-dancing clubs, where young women perform under the male gaze) has restored in them a sense of self-confidence, a reconnection with their physicality. This is, indeed, balancing on a fine line. In Chapter 4 I considered whether pole classes can be said to 'mend' something in some women's lives and it would seem, from these accounts, that it can. Some feminist scholars would argue that this is the result of women defining themselves only as sexualised beings, as lacking consciousness, and I am not trying to deny that either. But does that mean that we should not enjoy being sexual beings, should not enjoy feeling desirable? Isn't it a positive development if a woman over 50 has an improved body image because she has attended pole classes? Isn't that what feminism would want? This is a difficult area; even a minefield. Whether its effects are limited or transitory (something I return to in the next chapter) are not really my primary concern at this point; what concerns me is that there are positive effects, effects which can be seen to be far-reaching, in that they imbue women with a renewed sense of self (be that physically, sexually or emotionally) – just, in fact, as consciousness-raising (CR) or other groups did for women in the 1970s. I recognise this is a contentious claim and, of course, before an outcry starts, I acknowledge that CR groups were politicised and linked to wider campaigns to improve women's lives. But isn't the personal also political? I believe that it is. Pole classes are, after all, as I have argued in the preceding chapters, run by women for women, to deliver fitness in a social setting, and with a stated aim to 'empower women with confidence'.

The oldest questionnaire respondent was 54 years old and her reply to 'is there anything else you would like to add?' was 'pole dancing

,n't discriminate, age or size' (81/F54/UK). The oldest participants at I interviewed were Lilia and Tia (although actually neither were particularly old in that they were only 47 and 55 respectively); Lilia attended an exercise pole class, Tia attended a strippery pole class. Both were professional women, attractive, friendly, intelligent, funny and also mothers. I interviewed both of them during their lunch hours, on different days in the same park in Sydney, and their pole schools are perhaps two miles apart. They did not know each other.

> I have continued on for 2 years, in fact I think I am one of the oldest students there, that is, oldest as in age and oldest as in being there the longest. (Lilia)
>
> I think when you've got confidence that comes out in meeting people and I don't know, you don't want to think that once you reach a certain age that all of a sudden you don't feel attractive anymore. ... You see I'm the eldest in the class, probably the whole group and being the shortest in anything I do, how can you feel sexy about that when you've got these tall, long lanky girls and you think 'I'm the oldest and the shortest'. (Tia)

I do not know whether Tia or Lilia would have wanted to attend classes for women over 35 as both seemed to enjoy meeting and befriending women of all ages; although, they both mentioned being the oldest in the class more than once and it had been a source of anxiety for them. Gidget may have been more comfortable in such a class as she disclosed that she was occasionally aware of, and uncomfortable with, being over 40 in a roomful of women in their twenties, despite her superior gymnastic skills. The oldest participants that instructors told me about were 60 or over and both were Australian instructors:

> JENNIFER: Youngest is as young as 16 and as old as 67. ... A really fit 67 year old but a lot of young 20 and 30 year old put them to shame, she had a 6 pack she is really [... unclear], she is super fit. Oh she is fantastic. ... Now that she is like a lot older she is going out and doing all these crazy things like martial arts and pole dancing so she is coming back next term actually which commences next week.
>
> SH: So what would you say was your average sort of age?
>
> JENNIFER: It is funny, it started of as being between 18 to 35 but I feel that we are now sort of getting the 40 year olds that are coming in as well like the mums of the kids so. We have

noticed this term that we have got women that are a little bit older, they would be in their mid 40s to late 40s [and] have never done much exercise and now they have you know, started this.

SAM: The youngest was 18. The oldest lady I have taught is 80!
SH: 80?
SAM: And she was amazing. She was so much fun.
SH: Could she do it?
SAM: Yes, she obviously wasn't able to climb the pole and do things like that – she did do a spin though when her feet were actually off ... what we call carousel, when her feet were off the ground. This was amazing – I didn't think she would be able to do it, but she did it. She pretty much decided she was going to do it – I must say, my heart was in my mouth, thinking 'thank God she has signed a waiver'.

So, if as the data seems to suggest, pole classes can provide a positive experience for many of the women who choose to take them, literally women of all ages, does it actually 'empower' them? What do they think about the concept?

'Empowerment'?

I discussed concepts and issues related to empowerment in the Introduction and in Chapter 3. Here I return to it in order to explore more closely how the participants talked about the term and if they had thought it through – and what, if anything, it meant to them. The dictionary definition, which I did not include earlier, is included now to illustrate that most people are aware of the general meaning of a word or concept and use it accordingly. Dictionary.com defines it as 'to give power to authorize especially by official means; to invest with power, especially legal power; and to equip or supply with an ability'. Interestingly, there is a note with the definition which states that its meaning was originally only legal in the seventeenth century, it went on to also mean to enable or permit, and was taken up by the Civil Rights Movement, and later by the Women's Liberation Movement. For example, in her work on goth women, Amy Wilkins found that

> [a]lthough the women I encountered do not frequently use the term 'feminist', they draw on the language of feminism to describe the

> benefits of being a Goth. Specifically, they use the language of 'choice', 'objectification', and 'empowerment'. These discussions, however, focus almost exclusively on sexuality rather than on employment or family concerns. In part, this focus is logical given the demographics of the community: Many are in college or employed in starter jobs, and most have not married or had children.
>
> (Wilkins, 2004, p. 329)

The term is now used in self-help, pop psychology and even in politics. Perhaps the history of 'empowerment', from a very well-defined legal term to its current more cavalier use, explains some of the wariness we ascribe to it. Certainly, Lilia, who worked in a legal practice, dismissed the term in favour of the term 'liberating'. I am sure we all accept that knowing the meaning of a word doesn't necessarily mean that we then formulate a philosophy or justification of our use of it. Maybe we do if someone asks us to and gives us time to do so. This gap between meaning, use, understanding and actual daily parlance is the point at which many academics have sought to understand the usefulness of what can be something of a hot potato: 'empowerment' can be limited, partial and subjective, and this is a fact I have had to think about when exploring pole classes and their meaning with the participants. Some of them used it, unprompted, to indicate their joyfulness about the changes pole had wrought on their lives. For example:

> It is liberating, I find it ... empowering, confidence, yes, it gives you that, to me that would be, giving you control of your own sexual way that you see yourself and showing everyone how confident you can be. It is an empowering thing. (Tia)
>
> It is empowering, you feel like wow! Look at me! (Ruth)
>
> I know what people say about pole dancing and they turn their noses up but I feel empowered by it, to me it is an empowering thing that I do. (Carole)
>
> Well, it is an empowerment thing, for me, you know, for me, I leave, I walk tall; it is empowerment. (Keisha)

CHARLOTTE: I believe the classes empower women.
SH: In what way?
CHARLOTTE: All ways! Sexually, socially, in your personal confidence, your feelings of personal power. I have never done anything like it before.

> I love it – I love the fact that it is the exercise and you can laugh and have fun doing it, and you feel empowered, it just, it boosts your confidence. (Hannah)
>
> Pole dancing has evolved into different genres, exotic, empowering and pole fitness. There is a great deal of difference between them and as the student you should try them all. (www.verticaldance.com, accessed on 20 June 2009)

Several questionnaire respondents also mentioned the term, for example, giving the impression that empowerment is a word more commonly used in the US:[1]

> It is empowering. (6/F47/USA)
>
> It's not sleazy or trashing, it's empowering. (19/F20s/USA)
>
> Different, challenging, empowering for confidence, fabulous fun. (32/F40/UK)
>
> It is feminine empowerment. (41/F49/USA)
>
> Made me feel better about myself, very empowering. (62/F25/UK)

Others were more cautious and reflective and so wanted to think about the term quite warily. For example, in this extract from my interview with KT, we are discussing the term and its prevalence in interviews I had done two years before:

> I think in the context of pole dancing when they say 'empowering', it means doing something that perhaps other people are thinking that they shouldn't be doing, and that in itself is empowering. But I don't know. I don't say you get empowered in my class – I will say it builds confidence and fitness. I am not saying that other people don't, it is just the way I do it. (KT)

SH: This whole concept about ... you mentioned it, empowerment – women say they feel empowered. Is that a word that you have heard them use, or is that something that you have observed would you say?

JESS: It is definitely something I have observed, hugely ... I don't personally use the word 'empowerment', it is just the thing to say in pole world I think, it is one of those ...

SH: Like a buzz word?

JESS: Yes.

It isn't enough, though, to just say that empowerment cannot be used in this context if a participant sees it as appropriate to their mission statement or to their own experience:

> The idea is that we are empowering women with confidence, and we have to be approachable – and these women want to chat to you about their every day life and how they are feeling, and they need to be able to open up to you about how they feel. And sometimes they get upset, you know, what I said earlier – the girl who started crying because she had [achieved a move]. So how would certain people react to that? And we want our women to be empowered with confidence and not to feel that they have to perhaps not get upset in front of people and instructors. So finding the right instructor is really important. (Evie)

SH: Empowered ... a lot of people use that word, don't they, and people can be a bit funny about it. But I am quite interested in it, you know, I think, can there be something empowering about it ...?

SAM: I do.

SH: But in that sort of context ...?

SAM: I remember doing a pole dance for a guy I really liked, and watching his face did make me feel really empowered – I felt so strong, and 'yeah'. There wasn't anything that difficult or anything like that, but it is more the ... people, they're going [pulls an amazed face] ... 'oh my God', it does, yes.

SH: Do you think it is as much empowerment through sexuality, rather than the strength?

SAM: For me though it was the strength – it was like ... [flexes her bicep]. ... The thing they also enjoy is the fact that it is the exercise too, and they are having great fun – they are growing within themselves, in their confidence, and they are finding their bodies changing too. It is all empowerment.

GENEVIEVE: It's kind of like an empowerment class. ... I say 'yes, it's brilliant for fitness but it's not a fitness class, it's a dance class, you just happen to get more toned, more fit.

SH: Empowerment is an interesting word; some people use it, others don't. Do you use it?

GENEVIEVE: Not in class. But if I'm talking to someone about it, I do.

Genevieve has qualifications in physiology and has done some teacher training to improve her pole teaching. She defined herself as a feminist and discussed the different ways that people learn, explaining to me her pedagogy of teaching and a general philosophy of pole. For example, she said:

> That's one of the rules actually, you're not allowed to say [negative] things like that. You're not allowed to say 'oh my fat bum' ... oh, I don't like that because it kind of brings the whole group down. ... I never actually say that, I never say that [the word 'sexy']. You know, like walking around the pole, and if someone looks it I will say 'wow, look at you, look at yourself in the mirror'. ... I'll say to them 'look tall and confident'.

This echoes Alison's policy (in Chapter 3) about how to imbue women with confidence, that is, not by exhorting them to be sexy but by finding ways in which to make them feel they are already doing something right. Ariel Levy (2006, p. 200) argues that 'women's liberation and empowerment are terms feminists started using to talk about casting off limitations imposed upon women and demanding equality'. But many of the things that feminism fought for have resulted, along the way mutating and shifting, into what we now see before us: the Pill, legalised abortion, equal pay, the right not to marry, not to have children, to have a career, or not, all these fights were the starting points (and many of them have not reached their end point even now). So we see that a logical, if perhaps lamentable result, is striptease culture. Children don't always end up how their parents wanted them to be! From striptease culture emerged pole classes which then itself developed into different sorts of classes, run by and for different sorts of women. By women and for women. Should we be allowed to pick and choose who can do what, who is ideologically safe, who can be 'trusted' to do pole classes? One of Roach's (2007, p. 39) participants said, 'the choice to be girly has to be acceptable to feminists. It's sour grapes to demand choice for women and then not like what some of us choose'. Similarly, as Martha McCaughey (1997, p. 158) points out, 'some contemporary feminists, including socialist feminists, believe that women's culture, experience and practice can provide the basis for feminist opposition to destructive patriarchal ideologies. ... Those who do not wish to accuse women of

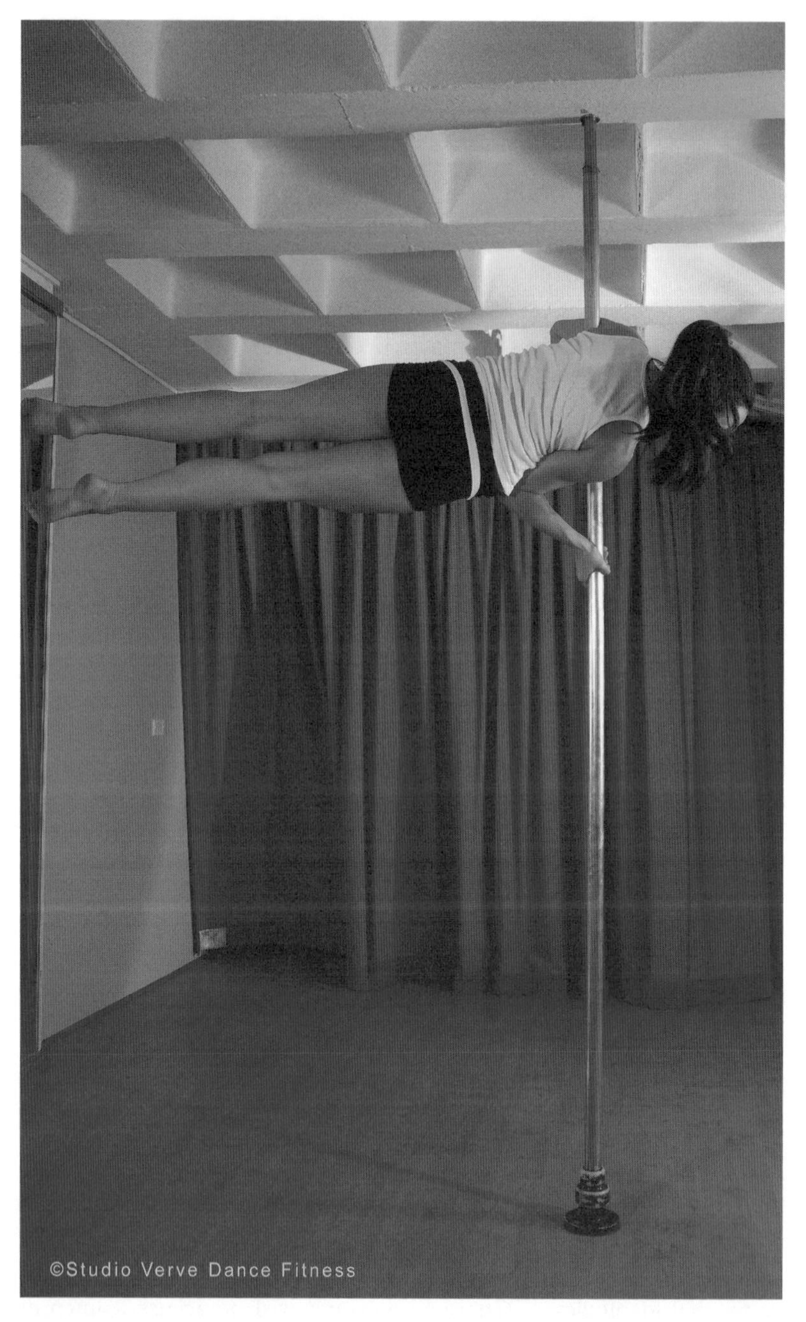

Figure 6.1 Jennifer Critelli doing the 'Flag'. Copyright: Studio Verve Dance Fitness.

being like men accuse women of being dupes of men. ... Those committed to a cause tend to understand those who do not agree with them as not acting consciously'. Whereas the women I met appeared to be aware of the contradictions around pole classes and striving to make sense of them, and to find cohesion and acceptance of pole as a valid, valuable physical activity.

'My classes empower women' – the pole v. the school

On 29 January 2009 a furore arose in the UK media about a pole instructor in Plymouth on the south coast of England and I include a brief account of it here because it illustrates several points about the current climate for pole classes. Sam Remmer, owner of the Art of Dance pole school, had been invited to a local college to perform two displays as part of a 'get healthy' initiative by the college. However, when she returned for the second of the displays the college told her that she had to move from the public area, where the first display had taken place, to a gym elsewhere in the building because of complaints from parents and teachers; somehow, over lunch, the display had been deemed inappropriate for its teenage audience. In addition, they asked her to not use any footage from the display despite their original agreement that she could. The row was first covered in the *Plymouth Herald* and then was included on the websites of *newlite.tv*, *BBC News*, *New York Daily News*, *Daily Mail*, *Daily Telegraph*, *Sun*, *Metro* and *Torquay Herald Express*. It was also reported on several UK television news channels including *ITN*. On her own blog the instructor (who I had not interviewed but who was, obviously, part of a wider network of other women I had met) dismisses the media attention as a storm in a teacup. To her the key issues were, and remained, the physical benefits of pole classes and encouraging people of all ages to enjoy exercise, which was the object of the display in the college. For example, she states that 'if anything my classes empower women and therefore encourage them to be in control of their bodies' (http://news.bbc.co.uk/1/hi/england/devon/7859239.stm, accessed on 23 March 2009) .

It is worth including the incident here, albeit in a much abbreviated form, if only because it illuminates for us that some people still associate the classes with lap dancing although these people appear to be no longer in a majority; that associating the two is a lazy way for journalists to fill some copy; that pole is still a subject which can stir up heated debate; despite that, many people now recognise the athletic nature of pole; and also because it gave me a window on the current attitude to pole. Some of the ensuing comments, taken from the *Plymouth Herald* website and the *Daily Mail* website (the latter particularly not known for its liberal readers) include the following:

> Oh dear! here we go again. I'm in my 70s and I feel so sad that there are so many paranoid and pathetic people in today's society who are constantly looking for hidden traits in all they see. This young woman could as easily have been seen performing as an Olympic display where she would have been applauded. Do please get a grip on the real world and see the beauty, not your so limited view of sleaze. (http://www.thisisplymouth.co.uk/news/Complaints-Plymouth-pole-dancer-performs-school-children/article-654900-detail/article.html)
>
> I am amazed that you can see nothing acrobatic about hanging upside down? Have you never seen Chinese pole in the circus??
>
> And as for not a fitness sport, you can burn up to 400 calories an hour... sounds like fitness to me!
> (http://www.thisisplymouth.co.uk/news/Complaints-Plymouth-pole-dancer-performs-school-children/article-654900-detail/article.html)
>
> I saw this on the news last night. It looks more like gymnastics to me! If she'd worn tracksuit bottoms and a long sleeve top, would there have been such a moan about it? Of course not ... it's a form of keep fit, unfortunately it's associated with those clubs! She should have used a long rope instead, then it would have caused no fuss!!
> (http://www.thisisplymouth.co.uk/news/Complaints-Plymouth-pole-dancer-performs-school-children/article-654900-detail/article.html)
>
> Personally don't find this offensive at all. I don't think that this is any different from a gymnastics floor routine. The dancer could have been a whole lot more provocative but as she isn't an erotic dancer or in a strip club she wasn't. Pole dancing is one of the best ways to tone up and get in shape. I'd like to see some of you whingers do some of the well thought out and demanding moves that this dancer has performed.
>
> Maybe you should climb down from your high horse and see this for what it is – gymnastics.
> (http://www.dailymail.co.uk/news/article-1131522/School-stages-lunchtime-pole-dancing-demonstration-pupils.html)
>
> If it encourages some of the lardy kids we have in the UK now to do some exercise, I'm all for it!

(http://www.dailymail.co.uk/news/article-1131522/School-stages-lunchtime-pole-dancing-demonstration-pupils.html)

I read the comments with a great deal of interest. I myself have probably thought, or at least heard, all of these comments and more in the past two years and had to try to make some sort of scholarly sense of it. What was particularly interesting was that the comments were primarily positive: writers argued that pole is a mix of athletics and dance, that the core strength and skill needed was impressive and that it reminded them of other gymnastic displays – ropes, horizontal bars, ribbons and hoops were all mentioned, and that if it encourages children to exercise regularly then all the better (as Kiwi Pole, who runs a weekly afternoon session for children who are accompanied by a parent, demonstrates). Those who commented positively also pointed out that pole as exercise was often judged negatively by people who didn't disassociate it from lap dancing clubs whereas it should be.[2] Of course, it is easy to see why there might be concern about pole classes for children: because of the sexualised nature of some of the linking moves. But, in general, pole moves are balletic and gymnastic in nature, and build strength and coordination; this seems to have begun to be acknowledged. This shift in public opinion was a revelation to me. On that note we move on to the next chapter where I focus on negative and positive aspects of pole classes.

7
'A Thing of Beauty'

I love how alive I feel!

(*25/F35/USA*)

The next two chapters function much as does taking a big breath between exertions. Chapters 4 to 6 focused on the classes, so this chapter discusses the positive and negative experiences of and outcomes of pole classes, a sort of midway rounding-up if you like. Carol Rambo et al. (2006, p. 224) argue that there are other narrative possibilities for a positive, active identity, where women enjoy their bodies, enjoy the attention and are aware of (and possibly sympathetic to) feminism; in other words, 'we own ourselves'. Conversely, Angela McRobbie (1997, p. 230) has noted the lack of 'active role models' which portray an active and energetic femininity. Arguably, pole classes provide those active role models. In previous chapters I have stated that, of course, pole classes are not going to achieve an organised, politicised movement which calls for improvements to women's lives within a patriarchal system, and nor would they claim to do so (as, indeed, neither would any gym or dance school). However, they are an organised movement and they do call for improvements to women's lives, however nominally: from the women who now have an improved, or hitherto absent, sense of positive body image; to those women who never thought they would enjoy exercise but now do; to those women who find the sociability of the lessons boosts their confidence and self-esteem; to those women who never thought they would end up running a business but now own a pole studio or school. Both Catherine M. Roach and Jacki Willson have pondered the complex and contradictory nature of sexualised, or previously sexualised activities, however positive, resulting in descriptive lists to best think through the dilemma, much as I do myself at various points in previous chapters. Willson (2008, p. 61), discussing burlesque, argues that it

> [p]ermits a more complex reading of representation, pleasure, self-image and the gaze. Undeniably the position is grey and ambiguous, and it seems impossible therefore to adhere convincingly to either of the stances (voyeuristic/exploitative/disempowering on the one hand and empowering/desiring/sensual on the other). Young art students are now at the point of wanting to explore this very emotive and complicated issue.

> What emerges from the stories for me is an ambiguous and complex picture of power, pleasure, agency and fantasy.
>
> (Roach, 2007, p. 26)

These lists – pleasure, self-image, the gaze, ambiguous, complex, power, agency – are all relevant to our examination of pole classes. So, in order to begin to attempt to unpick the lists apart, I now turn to the negatives about pole classes.

What are the negatives?

Previous chapters have mentioned or discussed various negative aspects of pole classes so here I return to them.

Bruises

Injury, as with any physical activity, is to be expected in pole classes although serious injury was not seen to be inevitable. Bruising was the particular injury common to everyone.[1] Some of the participants discussed bruises in their interview:

> People always say to me 'Oh will my bruises get any better, are they getting smaller?' and I say 'The only time your bruises will go is when you stop pushing yourself' because bruises come from poor technique; good technique comes from learning a move.
>
> (KT)

KT's website is also one of the many pole school websites which address the issue:

> Bruising and chaffing is generally caused by lack of body strength. The arms are not strong enough to support the bodies weigh so the legs

> over compensate by gripping on to the pole too tightly, alternatively the hands grip the pole too tightly allowing the lower body to spin but the lead hand to stop, this will result in chaffing of the wrists and arms. As your strength and technique begins to improve your bruising and chaffing becomes less and less. For wrists and arms we recommend placing sweat bands on the offending areas, for knees and legs, you should buy dance knee pad or use rolled over leg warmer and for feet you should try jazz shoes or yoga trainers.
>
> (http://www.verticaldance.com, accessed on 2 June 2009)

Suzie Q, James and Gidget also discussed bruising:

> I get a lot of people who haven't exercised before and they don't realise that strength based workouts hurt and they are going to be sore. Particularly with pole you have got the added things of bruising and that kind of thing. ... Girls come in the next week and compare who has got the best bruises ... and they don't really hurt, they just look impressive. ... It is just when you are burning, for want of a better expression, burning new bits of skin, so for new moves you bruise, then it stops. (Suzie Q)

GIDGET: [Husband] would say 'have some [painkillers] honey and you'll be fine, go to class' [laughter] because he knew that it made me so happy. So he was sort of like, 'no keep going', and then about after a year, I got over all the pain.

SH: What about the bruises?

GIDGET: Yes, the bruising wasn't such a problem, I mean, except when you first start.

SH: Did you have the body strength to hold yourself away, because it's about, you get less bruises as you get better don't you?

GIDGET: That's right. First, probably 4, months where you would get the bruises and you can tell what level someone is by where the bruises, oh yes, you're into [level] 3, you're doing that now.

SH: What like?

GIDGET: Oh yes, mainly well when you first start, all up your shin bone and then as you get on, up higher. I had really big ones on the inner thighs as I advanced, I've had one on the back of my hips and my elbows – and you get a random one and think what are you doing with that?

SH: I still get this [indicates top of foot], here, it really hurts there.[2]

GIDGET: Yes, I've had them all and then you just seem to get over that, once you get past that pain you just keep going and that's when I just thought, I'm 40, I'm still here and I'm not hurting any more, so I must be all right.

JAMES: I have to say I got off the pole [after his first public performance], and I always have a drink after it and my beer was on the counter, and I couldn't pick it up – my lower arms were that tight I actually physically couldn't grab on because it was so intense.

SH: Yes! Did you bruise at all?

JAMES: Yes I did but not too much but some of the guys that I have seen have the biggest bruises here [indicates legs]; it is just part of what it is.

44 questionnaire participants mentioned injuries; primarily bruising, but also pole burn, aches and pain and other minor injuries (only one said she had twisted her ankle).

Cost

The cost of pole classes, and how women 'find' the money to attend, is something I wrote about at more length elsewhere (Holland, 2009, pp. 40–1) and I do not wish to reiterate it here. However, briefly, I explain the difference in cost between a typical gym or fitness class and pole classes:

> Exercise classes at a local council-run leisure centre cost £3 at off-peak times and £5 each at peak times (although pole classes are not available, these are the prices for classes such as aerobics or pilates) which means that a seven-week course would cost a minimum of £21 or a maximum of £35. In contrast, the pole exercise classes cost £20 for an initial two-hour 'taster' session and then £135 for the following six weeks, a total of £155 and a difference of over £100 (£100 equating to about US$190 or AU$250, at the time of writing [2008]). Of the 15 women interviewed for [stage 1 of the] study, most said that the cost was high but because of the 'difference' (by which they meant novelty, the speciality) of the activity they did not see it as a problem.
>
> (Holland, 2009, p. 40)

Seven questionnaire respondents said that they found classes expensive but intended to continue attendance anyway. In general, the attitude of the participants echoed Lilia and Charlotte's rationalisations of the cost:

> I don't have a lot of expendable income but I don't think it is expensive because it is extremely good value, you get very very fit, you have a lot of fun and it is interesting. (Lilia)
>
> It's kind of expensive, perhaps, but you know, you get what you pay for, that's what they say isn't it, and there are so many benefits with it, well, I think it ends up being worth the dollars. (Charlotte)

This type of rationalisation of the cost was common with participants listing the benefits as a kind of mantra to justify their continued attendance.

Perceptions of pole

This was the key negative aspect of pole classes and quite clearly a refutation, even a negation, of their claims to 'own ourselves'. On the questionnaire, as I note in Chapter 2, I asked, 'what are the negative things about pole classes?' to which 22 respondents replied that attitudes to pole annoyed them, where people assume it is 'sleazy', 'tarty' and stigmatised. The media is blamed for many of the incorrect perceptions, as are sheer ignorance, lazy thinking and 'uptight feminists who don't even know what a class looks like' (Grace). Evie said that pole dancing still has a stigma:

> I kind of felt that I had the need to explain each time, when people were making a sort of enquiry ... even so now with Pole-da-Cise; people say 'oh I am phoning up about the pole dancing'. And I kind of feel like I want to say to them 'well, we are Pole-da-Cise, which is pole dance exercise – and a form of pole dancing, I guess'. But there is still that stigma out there that a lot of people feel that if you are a pole dancer, the chances are that you are working in a nightclub and you take your clothes off dancing around the pole and you get paid for it. And I have got nothing against pole dancers whatsoever, but that wasn't what we wanted to be about.

Suzie Q studied public relations at university:

> I wrote my thesis [about] the fact that because of the industry I am in, even though I excel and am recognised as one of the best in Australia, I have no respect from society. And it is changing but I am

never going to stop meeting people who go 'oh, pole dancing' and yet they don't fully understand what is involved in it ... their ideas of pole dancing just come from the media portrayal which are, you know, obviously not brilliant.

The participants of both stages of the research also discussed the issue:

I think, as I say, men have difficulty getting their heads round the idea that women are getting something out of it that isn't connected with looking at nakedness, and they are just not looking – they are just not seeing it from a point of view of doing it at all. So it is hard for them to get their head round, in one way, why women want to do it. But if they sit down and think about it and they think 'if my partner was doing this, what would that do for us?' It has got to be a fun thing, it has got to be a good thing. No one is breaking any laws, no one is taking any clothes off, no one is offending anybody ... I do feel irreproachable, I do. (Rachel)

Pole dancing is able to help with [body image issues], maybe because it is an all-female environment, I don't know. That is what else annoys me, they talk about it being exploitative or whatever and overly sexual, sexy – and, if you are a woman and you have never done any exercise and you are kind of [makes a straining noise], it is not sexy. ... But there are no men there! So who is being exploited, who is supposedly watching this? (Suzie Q)

I do get sick of people thinking I am some sort of whore, or an idiot, when I am neither. (Grace)

Despite the classes (with very few exceptions) being all-women, several participants reported that their male partners were uncomfortable, for example, as Silke discussed:

SH: Has he seen you on a pole?

SILKE: When we did the fun classes, yes, but obviously not since, no. But my friends know how important it was to me at the time and how good it made me feel, and you know, whereas he just ... he doesn't really want to discuss it, *I don't think* [my emphasis]. He was shocked when I told him – I did tell him I did it last year, just so if ever it came out that I hadn't lied to him. And I was shocked that he was shocked, because I thought 'well, how well do you actually know me? Because if you knew me ...', I mean my friends weren't surprised

> whatsoever. They were like 'oh good, it is about time you joined that', you know.

Silke had not tried to discuss her partner's discomfort with him and so was unclear about his reasons; neither had she attempted to explain to him what a class was actually like. Hen parties,[3] which is where many pole instructors, including Silke, Rachel and Gidget, started to teach, are usually held on an afternoon or early evening and involve a group of women, some poles, some tuition and often alcohol and food (which don't necessarily mix well with pole work). Pole parties are designed to be fun. Several participants suggested that events like hen parties were useful for giving pole classes a new image, for example Alison said that

> [hen parties] break those stereotypes and break that taboo and I think that is really helpful in spreading the word. ... A lot of people have heard of [classes] because of hen parties and that's quite a useful tool to breaking [negative] perceptions.

The negative aspects of pole classes, then, can be seen to be not surprising (as bruising and soreness are common with any sort of physical activity, and most activities cost money), and very surprising, if you consider pole to be an athletic pursuit. Of course various dance forms have attracted disapproval throughout history, from ballet to salsa, so perhaps pole's current status as a 'dangerous' art form will be eroded as pole continues to evolve and prove itself to be both athletic and balletic; both the same as that performed in lap-dancing clubs but also moving away from it. I return to perceptions of pole in the final chapter.

What are the positives?

Pole classes, according to the participants of this study, provide a positive experience for the women who choose to attend, for a variety of reasons. The Pole Stars website uses feminist rhetoric to market the classes claiming that they are 'safe, friendly spaces' with a 'policy of inclusion' – which, as I pointed out in Chapter 4, isn't strictly true considering the classes are expensive and held in a pub. But as others (for example, Tara Brabazon, 2006) have argued, exercise is, on the whole, exclusionary because many people simply cannot afford to attend classes or join a gym. Michelle Segar et al. (2006) found that a social reason is one of the best ways to ensure continuing an exercise class or routine. Indeed, women who exercise only in order to change their body shape or size rarely exercise as much, or enjoy it as much, as those for whom the motivation is primarily social. The latter are also more likely not to drop out of the class. Lisa McDermott (2004) examined women-only canoeing trips, finding that the support and camaraderie for the women who went on them outweighed any concerns they had initially had about whether they were able to do it. Again, there are issues of cultural capital here as only women who can afford the time and money to go on such trips could accrue the benefits of them.

Music

The very fact that pole classes include music, of all genres, made it popular with many of the participants, for example, Evie explained how playing the favourite music of two very overweight, shy potential students encouraged them to try out classes:

> SH: You said that you were playing their own favourite sort of music?
>
> EVIE: Yes, we spoke to them and said 'what sort of music do you like', and they had both come along and said 'we love dancing – we love getting out ... we don't get out that often, but we love dancing'. So I said 'what sort of music do you like?' So they liked R&B type music. So we play, in particular I made sure that every time these girls were in for their lesson, we always have different types of R&B music.

> I just used to go and I'd meet all different ladies and the music was great, and I was away from the children, like that was my little thing. I think I was 38 when I first started, no I was 39. ... [The studio] is all pink and the lights are pink, so when you look at yourself in the mirror you look fantastic. You start thinking, oh, I'm not bad at this, I should keep going ... [I felt] toned up and just felt really good about myself and the big thing for me was the music, just maybe it brought back memories of dancing and gymnastics, but hearing the music and being able to move, I can go to a pole studio and dance and have no one watch me ever, just dancing to that music is enough for me or dancing with a friend is enough for me. (Gidget)
>
> It's good to hear music, loud music, I liked that. (Jane)
>
> I like the vibe, the music, the voices, it makes it, you know, a thing, it makes it exciting. (Jessica)

Similarly, Evie related how she had missed dancing after her children were born:

> [Dancing] was kind of something that I enjoyed a lot. And then when I, as I got older and then I went on to have children, you know, those kind of things are sort of left behind a little bit, aren't they, because your life takes a different track. So although it was something I still liked, it wasn't something that I did so much, like most people when they sort of have a family. ... I missed it tremendously because it was part of me and it was like being young. So there were times that I, much as I loved my husband and my children, there were times when I thought 'well that was part of me only', and it sort of got left behind a little bit, you know, it is no longer ... 'I can't be selfish and think about me all the time – I have got to think about the rest of my family'. So yes, I did miss it.
>
> (Evie)

Listening to music as a leisure activity, or going out dancing, are activities commonly curtailed as women grow older and/or have children and pole classes (as do other exercise classes to music) provide a place to experience the physicality and the opportunity to listen to music without interruption. Pole classes particularly seem to offer women a range and choice of music.

Dressing up

For 'high church' or strippery participants, a positive aspect of the classes was the opportunity to dress up in stripper heels and various outfits, as discussed in Chapter 5. In fact, this aspect of classes fits in with other work about women and sexuality:

> [The goth] scene's celebration of active sexuality as resources to resist mainstream notions of passive femininity. ... Contemporary young women in a variety of arenas use active sexuality to stake out gender independence. This emphasis on women's emancipated sexuality reflects the substantive turn of postfeminism – what Anna Quindlen has labeled 'babe feminism' (1996, 4) – a focus on women's right to active sexuality rather than on broader issues of gender equality.
>
> (Wilkins, 2008, pp. 328–9)

As we see, this reflects some of the same issues as strippery classes with its stress on active sexuality, individual gain and a lack of politicisation while simultaneously drawing on feminist gains and rhetoric. As Elizabeth Wilson argues in *Adorned in Dreams* (1985), dressing up can also be pleasurable. I return to this in the final chapter.

Friendships

As Eileen Green (1998) has noted, friendships and networks are crucial for women's sense of self and well-being. A large part of the appeal of the pole classes was the very fact that the classes were all-women and that there were no male observers, what Tara Brabazon (2006, p. 67) calls 'an opportunity to create feminine autonomy and space'. This was not apparent in the first week or two, but as the classes progressed, and the students became more used to each other and found themselves enjoying their own efforts and encouraging others, the sense of camaraderie became integral to the overall experience.

> There is a lot of clapping goes on, and the people who have been coming for a little while, they know how difficult it was when they started, so they are really encouraging to the others. And I think it is basically that which gives you your confidence: that people say you look good, and eventually, after enough times, you start to believe them. And when you believe them it gets better because you have got the confidence.
>
> (Rachel)

Glass (1999, p. 58) recommends that 'the atmosphere should be comfortable and encouraging not intimidating or competitive' to encourage women to exercise. In stage 1 of the research, the class was divided into two; each group sharing one pole and working in pairs, so one might expect a feeling of rivalry between the two groups as well as between the individual women. However, there was no competitiveness between students as several women pointed out; for example, Carrie said:

> We weren't in competition with each other, you were glad if someone worked something out, you know if they worked out a move. I mean, it made you all the more determined to have another go yourself but in a good way, like you were glad she had done it, she had shown you it could be done, so you were happy for her and for yourself.

Their focus was primarily one of exuberance about enjoying exercise (a new feeling), of feeling that they had progressed, however limitedly, and the added, and unexpected, bonus of the bond with the other

Figure 7.1 Students and teachers at a Purity pole jam. Photography: JasonParlour.com.

women in the class. Like the male dancers in Gard's (2003, p. 114) research, the participants of this study 'return frequently to words such as "passion" [and] "freedom" ... intense words which suggested a kind of joyful abandon'. There was no indication in the interviews that the participants expected this type of outcome, but certainly freedom or joy or abandon were frequently reported or inferred as being a large part of their experience of pole classes. In this way pole had an appropriately 'feminine' approach, particularly for those who were not athletic, because it lacked a competitive ethos so echoing received ideas about competition being 'masculine'. Evie explained how encouragement was the norm rather than competition:

> We have a class of 3 who have been working together for probably 10 weeks, maybe a little bit more – and we introduced a new lady into that class who has probably done 6 weeks. So she was a little bit behind – but reading from my notes, I felt it would benefit her because she does pick things up quite quickly and she is quite strong. ... However when the girls individually take their work onto a crash mat, overseen by me, and they go into their move – [she] was doing half of that move, if you like. So it wasn't as if ... she could see where it was leading, and so watching the others would help her progress – and

> they were very much encouraging her and clapping her and saying to her 'I never did it that quick!'.

This much can perhaps be expected of a feminine environment such as an all-women class. However, the pole classes did not fit into other feminised styles of response, such as those found by Maxine Leeds Craig and Rita Liberti (2007, p. 890):

> Women who wished to fit in at GetFit adapted to a feminized interactional style of nonjudgmental and noncompetitive sociability ... in which they praised other members, downplayed their own achievements, and signaled their own low status in relation to exercise and conventional norms of beauty through self-deprecating humor. ... Women bonded in the gym around their aversion to exercise rather than any pleasure they might find in the exertion or feelings of pride they might take in their competence.

Certainly, the polers were non-judgemental and, to a point, non-competitive; the classes were sociable; and humour was used, which I return to, later. However, despite some polers having had a previous aversion to exercise, they did not bond around their own incompetence, they did not downplay their achievements; they were not apologetic about their own, or others', achievements.

During stage 1 of the research, part of the lesson was to learn 'filler' moves which didn't involve swinging round or hanging off or climbing up the pole, and which the teacher described as 'sexy', so we were told to do, for example, little shimmies or some simple footwork. While most students took everything they were taught seriously, listening and watching intently as the teacher explained and then demonstrated different moves, the 'fillers' were the thing that prompted ribald teasing and reluctance as well as encouragement and assistance. The 'fillers' were initially what caused the most self-consciousness but were also the reason for the most of the laughter. Kandy James (2000, p. 265) argues that 'embarrassment can inhibit participation in any encounters, including those in leisure situations' but contrarily it was the embarrassment that forged stronger links between students. Teela Sanders (2004, p. 274) found that female sex workers 'manage their work ... through joking relations' and the use of humour in a class such as pole also serves to diffuse the possibility of the situation becoming sexualised; pole was not only a feminised space but also a heteronormative environment. Encouragement and admiration of other women when they are striving to do pole moves was one thing. Doing sexualised filler moves seriously, in the first few weeks at least, in a roomful of scantily clad women was quite another.

> With Jen's class it is the most loving feminine supportive environment that I have ever come across. ... Everyone looks after each other, everyone says positive things about each other, we have a huge amount of fun, we laugh like lunatics most of the time, and it is just wonderful. (Lilia)
>
> I've made some really good friends and I know on a certain night they're going to be there, so that's exciting, so you go there and do this routine together and its just, especially if Bobbi dances as well, it's just, something magical about that. I don't know what it is, no one has to be watching, just that doing it, you get something out of it. (Gidget)
>
> Half the time, we couldn't stop laughing, we just couldn't, so we couldn't breathe and then you can't even do anything so you laugh even more, it was stupid really, I just loved it. (Grace)

As Eileen Green (1998, p. 171) suggests, 'in particular circumstances, women use humour to subvert sexist imagery. Shared humour between women in leisure contexts, can be a source of empowerment and resistance to gender stereotypes' and this was certainly the case, although it was not self-deprecating humour.

Instructors

Maxine Leeds Craig and Rita Liberti (2007, p. 688) found that the employees at the gym they studied provided 'unwavering, friendly, and nonjudgmental support for customers [which] constituted a form of emotional labor'. One of the most interesting themes which arose from the data was the inspiration and confidence that the instructors instilled in their students:

> I am very careful and Jen is very careful about what you do, she won't let you do things that are going to endanger you, and so she said 'I can get you to level 5' and I used to say 'you silly thing, get over yourself'. And she did slowly but surely, she has egged me on, and I would get into my comfort zone and I would want to stay in a particular class [Lilia did the level 1 class three times] and she would go 'no, time for you to move up'. So yes, it has been really good, and it is fantastic as you know, it is just wonderful, absolutely wonderful ... it was just so much fun, it was just ridiculous fun. (Lilia)

> Our teacher was really good, really made me feel confident, even when I was telling her I could never do it. She would just look at me and say 'come on, think about it' and I would think, well I was thinking about it, then I realised I was only ever thinking I couldn't do it, so I started to think I could do it, and then – well, you know, I started being able to do it. It was good. But without her, I wouldn't have done anything. (Tij)
>
> She was very patient. I sometimes felt so stupid, at first, really, you know, dumb, and heavy, like a big old cow, and she would say 'well, do this and then this' and I would, straight away, it would go, I wouldn't remember what she had just shown us. But she would never get pissed off. And I started to realise that she wasn't going to shout at me, so I felt ok, I was feeling good, I started to remember. (Jane)
>
> I mean it took me, I would say, 4 or 5 months to get a foot off the floor. But because she is who she is, and she was my friend as well – and to be honest I don't think it was because she was my friend, I think I would have done it even if she wasn't; she is one of these people that inspire. (Libby)

Most of the instructors have transformative stories to tell about women in their classes. For example:

> If you've got a good teacher who can understand your limitations and push you in the right direction then you'll definitely leave that class having achieved something, even the littlest thing is an achievement you know. (KT)

> I think in all the teachers that I selected, I selected them not on how they looked, but how they actually were. Like, I had teachers from a yoga background, dance background, gym background – every different kind of background. But it was more how they would understand the person and how they would talk to them, made sure that they didn't talk down to people or try and show off. ... I mean the most important thing is breaking the move down and building the students up.
>
> (Sam)

Sam explained how the classes could benefit both instructors and students alike:

> They will often have like an idea stuck in their head that you have to be skinny, tiny, with long blonde hair down to your arse, and

> then they will come and see me standing there in shorts and T-shirt and bare feet – I am an average sized woman, and you know, when you actually show them that they can do things, they sort of trust you ... I got so much out of doing it. ... And I am a lot better now at talking to people and giving my opinion on that type of stuff – but the one thing that I really found, as a teacher it is astounding how the students would want to come up and talk to you about, not just pole problems, but problems outside in their life and all the rest of it. And women are really hard on themselves. It is like a trust thing they have, because you have taught them how to be sexy again and that type of thing. They want to share things.

The relationship between instructor and student was mentioned by 57 of the questionnaire respondents as being an important and positive factor of the classes' popularity. It seems that the pole instructor is somehow more than just the person who runs the exercise class; instructors can be seen to be the reason that women feel the positive effects of a class, as well as actually being why the class is there in the first place. So for the participants to feel friendship or affection, and gratitude, is perhaps not surprising. What is more surprising is to find that, in something as seemingly problematic as a pole class, strong and active women are running classes and performing emotional labour, and students are experiencing them as both role model and as friend. I return to this in Chapter 9.

Toning not bulking

Muscular endurance and coordination are aspects of pole perhaps least likely to occur to people, as opposed to nakedness and titillation which are often the first things that occur to people. As the Pole Stars website explains, pole is not just about swinging from a pole:

> Imagine having to support your entire weight with one arm, or supporting your upper body weight with your stomach muscles to get an idea of the level of effort involved.
>
> (www.polestars.net, on accessed 22 February 2008)

So, while the women were aware of the erotic connotations of doing pole, they were also aware of the physical benefits and, for exercise polers at least, preferred to concentrate on those – despite the non-exercisers' loathing of the mat-work exercises that would improve the physical benefits (discussed in Chapter 3). Clearly the idea of hidden strength was appealing as it does not 'masculinise' the woman in the same way that body building does (Mansfield and McGinn, 1993). I discuss elsewhere their admiration of the hidden strength it takes to do pole successfully (see Holland and Attwood, 2009, p. 176), and had pole involved 'bulking up' it would appeal to different people than it currently does. Sarah Grogan et al. (2004, p. 49) found that women 'are as scared of becoming muscular as they are of becoming fat' and while some of the students told me they could see the difference in their muscle tone, particularly upper arms (for example, Suzie Q said, 'it gets rid of those bingo wings' and Keisha said, 'I just wanted to feel my arms all the time!'), they were not concerned because they were not beginning to bulk up.

> I see so many students but I definitely, rather than physical unless it is huge physical change I notice strength, that is what I tend to focus on more because like I have had girls who couldn't even hold their body weight on level on 1 and they have now made it to level 6 and they are doing things that they never thought of. Like I had a girl who sat on the floor with her legs out in front and she was so tired that her legs had bent like that she can't actually extend them and sit up straight and now she can so and that is not even from doing any extra stretching or anything that is just coming to one class a week and the body getting adapting to moving a different way and the muscles lengthening out a little, so that kind of stuff I notice.
>
> (Jennifer)

38.5% (52) of the questionnaire respondents said their body image was 'OK, (could be worse)'. 21.5% said their body image was 'excellent (I look good)' and 37.8% (51) said their body image was 'up and down' (they had good and bad days). Only 5.2% (7 people) said their body image was 'generally bad (I hate my body)'.

Achievement

Over and over again in the questionnaire responses the respondents write about feelings of challenge and achievement, when 'nailing' or 'busting' a move, getting fitter, managing to keep going to class, meeting other people and generally feeling a sense of increased confidence.

LILIA: I am fitter and more liberated in my body, it just feels really good ... I feel amazed that I can achieve physical outcomes that I never thought I could possibly do.

SH: So you obviously didn't get these same benefits from aerobics?

LILIA: No, no. ... [Aerobics is] mind numbing, there is really nothing going on there, it is incredibly boring. With this you are actually achieving something and I have never been able, you know, I could never climb on the monkey bars so to climb to the ceiling on a pole, I think it is a huge achievement. You put yourself upside down, hang off with one leg, it's a big achievement, it's hard. ... It is an amazing success and it does look incredibly pretty.

> I think it probably is – with the gym, you do the same thing. ... But with pole dancing, you know, you are going upside down, you are climbing the pole and then going upside down. And then you are like climbing the pole going upside down and coming down really slowly like a panther ... things like that, it is all building on. But you are actually giving things you can show people. (Sam)

One theme that was particularly strong in all the interviews was the admiration all the women felt for pole dancing itself. They saw it as an art form, as a difficult and athletic skill and as something elegant to view. But none of them saw it as inherently 'sexy' or titillating, rather they were aware of the strength required and so were more admiring of its beauty. Rachel argued that

> men can't understand, because they see it as a purely ... it is a titillating thing for men, that they can't even be bothered looking outside that box. They just can't see that a woman, a heterosexual woman can get any benefit from seeing a woman pole dance. They can't see ... they can't put themselves in a different mindset to see that she is thinking 'I could do that – I could make myself feel good, I could feel powerful up there'. ... It is a thing of beauty ... it is pure enjoyment.

Silke explained that she loved watching other pole dancers:

> I think it is amazing to look at, really amazing, when you think that someone is holding herself up there like that, and she can do all those things, just by using her muscles and a vertical pole. It's very

> sad really that, you know, that people can't get past it being about stripping and supposedly being a spectacle for men's eyes only. It really should be seen as more than that.

Another teacher, Chrissy, also mentioned the physical strength needed to do pole exercise:

> Pole is so athletic and balletic. It takes power to do it, you need balance and control and strength and grace, and you need confidence as well, swinging, climbing, you are upside-down, you are hanging off there and no one but you is doing it, you are using your own power.

Students echoed the sentiments about the aesthetic appeal of pole exercise:

> It is very beautiful to watch, a thing of beauty. I think that now even more so cos I know how much strength they need, it's like swans with all the beauty on top, on show, and all the paddling, the strength, going on underneath. (Jane)
>
> I love to watch it, I never lose the, you know, I am still awed by it. (Kate)

> LIZZIE: It is just brilliant to look at, don't you think?
> SH: Yes, you think 'how is she keeping up there?'
> LIZZIE: That's it! It's like you stay up there and it's 'look mum! No hands!'

This image of women as strong and powerful, viewing pole dancers with admiration, contradicts the image of pole dancing as only titillating for heterosexual men.

Improved confidence

In previous chapters I noted that many of the participants of this study reported that they felt an improvement in their overall confidence as a result of attending pole classes, which is an issue appropriate to briefly return to here. For example, both KT and Jess immediately agreed that pole seemed to have an inspiring effect on many students:

> KT: That's it exactly but yes I mean, people totally change, we've had people that will not say boo to a goose like, and it's incredible

and then all of a sudden they're all strutting round and it's a hugely social thing pole dancing classes as well, and I've never experienced that like in aerobics or in a team sport or anything like that, where people really seem to sort of bond I think because, I don't know why, because it's ...

SH: So do you think the effect of a pole class can last outside the class?

KT: Absolutely, I think they're inspiring, instrumental in making people exercise more. Because they want to achieve more in their class, they know that they have to do more outside of their class to improve their fitness levels and also flexibility because like stretching when they wouldn't normally.

JESS: This girl came in once and it was early on when I started teaching – she came in and she was so shy, but she was on her own which amazed me, because this shy girl came to a pole class on her own.

SH: Do most people come in pairs?

JESS: A lot of people come in groups, so 5 girls will book onto a course or they will come as a group, or there are some people that come on their own. But this lady came on her own and she was so shy and she was 'no, no, I am not going to try that, there is no way I will do that', and now she comes once a week, she is upside down, she is spinning around, she is loving it. ... She said 'I feel so confident about myself now'.

LILIA: I wanted my daughter [age 16] to do it as well and she was like 'oh ma, I am not coming to pole dancing' and her father was like 'oh you can't do that' and so I took her to a show to watch them and when I got there all the girls came over and we gave each other hugs and kisses and my daughter said 'you have some nice friends, they are young and interesting, why do they like you?' She was like 'why are these women so nice to each other, it is this totally wonderful environment', and of course now my daughter is totally addicted and goes as well ... sometimes I do a class with her and it is great, everyone is nice and supportive.

SH: Did your daughter enjoy exercise before?

LILIA: No. She did nothing at all. When she was little she would ride, when she was very tiny she did a gymnastics class but for years she hadn't done anything. I was concerned about her

> not exercising ... but she was very hesitant, well, all of my kic thought it was disgusting. My oldest son wouldn't, he would not even listen, he would put his hands over his ears and say 'I do not want to hear this'. I think he thought my mother is doing pole dancing, he had visions of me in a g-string and high heels! ... But over the years they have gotten used to it. ... It is quite normal, I have photos at home.

Lilia found that pole classes forged a new bond with her daughter, at an age when daughters and mothers are usually moving apart from each other. The classes that Lilia attends are not strippery, which may have made her feel more able to suggest that her daughter begin to attend. Many polers believed that the benefits of pole were lasting, and were felt outside the class as well as during it, as both Tia and Alison's responses illustrate:

> SH: So you said you were slightly shorter you know, so can you, as you look back, can you see how its affected your body confidence?
>
> TIA: Oh definitely, to me I wear slimmer type clothes, I wear dresses, I never used to wear dresses for years, I would wear trousers, I was getting sick of that and then I thought, I don't suit skirts and then my daughter said to me one day, she said, 'try a skirt and see what it's like' and so for the last couple of years I've been wearing skirts and more fitted dresses and high heels and whatever there is.

> It actually upsets me when people call it a 'fad' because it really isn't, there is so much to it ... I think that's why it will last, because there are real concrete benefits, real benefits and a real impact. (Alison)

In this chapter the positives outweighed the negatives which is a direct consequence of what is contained in the data; issues about embodiment and body image, will both be discussed again in Chapter 9 and in the final chapter. This chapter was meant as a recap, before the next chapter widens our lens to examine the many ways that the global pole community has grown, thanks mainly to the Internet. The two chapters after that then narrow the focus to two case studies, revisiting some of the issues touched upon in this chapter.

Pole Community: Opening Closed Minds

Much has been written about 'remote relationships' which are conducted at a distance, primarily through technology such as the Internet, for example, Smith and Kollock, 1999 on communities in cyberspace; Bruns and Jacobs, 2006 on blogs; Whitty and Carr, 2006 on online romances; Holland and Harpin, 2008 on teens and MySpace. There is also a great deal of research about women and their remote, gendered relationships: for example, Spender, 1995, on women and talk online; Cherny and Weise, 1996 on women and communities online; Consaluo and Paasonen, 2002, on women, agency and identity online; Thiel Stern, 2007, on girls and instant messaging; and Bury, 2005, 2008 on female fans online. Among all this scholarship runs a common theme: that the Internet suits women because they can be sociable when they choose, when their children are in bed, they don't have to risk going out at night alone, or when they don't feel like having to dress to 'go out'; and because online communities are, to some extent, cooperative and communicative, which, traditionally and theoretically, suits the way women interact with others. The Internet in particular, but also other technologies such as the mobile/cell phone, have been liberating for women and certainly, in this chapter, we see how the increased popularity and availability of the Internet has been of benefit to polers worldwide.

As I argued in the previous chapter, pole is an organised movement with its stated aims generally being to improve women's lives by offering increased fitness, confidence, friendship and fun ('fun' being a word used repeatedly in both the interviews and the questionnaire). But polers are aware of the image that some people have of pole dancing, even of pole classes, and so strive to find ways to improve its profile and its image. Genevieve told me that 'the pole dance community is so together and really, really supportive of each other' which, overall,

reflected what all the other participants and questionn
reported. There were one or two exceptions. One inst
polers became more advanced the level of competiti
ness became much more pronounced 'until it gets a
anyone will stab you in the back'. Similarly, KT discussed at len
she attempts to manage in-fighting on her Vertical Dance discussion
forum:

> I am very strict on it, and in starting up this business, the hardest thing I have ever done was open up that forum. And in hindsight I still probably would have opened it, but it has caused so many arguments, so many fallings out, because you can't make 100% of people happy 100% of the time. And I was very, very strict in the fact that I don't want swearing on it, I don't want people's personal digs about other people. I don't want people naming and shaming people, because at the end of the day I would like to think that if I was in the person's position getting named and shamed and it wasn't true. And I don't allow ... it is for pole dancing and should only be for pole dancing. And if you want to go and talk about whatever, go somewhere else. And that has caused ... I have had to pull people up and they have got angry, and I have had to ban them because they have gone and been really horrible. I try to keep it as less bitchy as possible, but it is very, very difficult.
>
> (KT)

In general though, all the students and most of the instructors did not report experiences of back-stabbing or arguments; mainly, polers want to believe in, and work hard to achieve, a supportive and welcoming community.

Online community

Obviously the Internet has made a vast difference to the ability of polers to communicate, no matter where they are. It has benefited the pole community in that polers can communicate across continents and time zones. There is a proliferation of message boards and discussion forums, websites and email lists, as well as Facebook, MySpace and YouTube, all of which are used by polers. The Internet, as I noted above, suits women because it can be fitted around with other things in their lives such as caring for family members. As KT said, 'YouTube is a good thing' because it is a way of publicising performances, competitions, skills and novations; KT filmed herself explaining the difference between pole as itness and pole as exotic dancing, in an effort to change the perception of pole classes and performance. Most school websites have a discussion forum; most have email lists to keep members up to date with new workshops and events. One example is the Pole Dance Community website (http://www.poledancecommunity.co.uk/, accessed on 12 July 2009), set up by Sam Remmer of Art of Dance in Plymouth (who is also mentioned in Chapter 6), which is 'dedicated to uniting the pole community'. It has pages for FAQs, news, approved schools and videos of performances by pole celebrities. Many polers on Facebook have a high number of mutual 'friends'. Polers can look at photographs of other polers, can comment on their moves, can see photos or videos of events and forge remote relationships which, in turn, contribute to the global pole community becoming more strong and cohesive. Many polers are online every day, talking to other polers, innovating and evolving their approach.

Olympics 2012

The pole community has a campaign to include pole as a test sport in the Olympics; the aim is to 'Get Pole Dancing/Pole Fitness Recognized As A Legitimate Athletic Sport'. The campaign can be found on Facebook and other pole websites as well as in publications such as *Pole2Pole* and *Pole Dance International* magazines, where a link to an online petition is provided. One Facebook group had 127,869 members; another had 704 (in July 2009). The campaign is at an early stage but fits into the model, discussed below, of ways in which pole (as with stripping in the past) is being mainstreamed away from its original image.

Pole jams

Pole jams were not mentioned in stage 1 of the research but in stage 2 several UK instructors mentioned them and they are often arranged on Facebook and via email lists. Genevieve explained what a pole jam is:

> What's happened as well is these things called Pole Jams, where, they started last year [2007]. So people are now, not only are they meeting in the [online] forums, they're actually physically meeting up and what we do is, we take it in turns, because there are more of us getting studios, you see. Dispensing with the idea of hiring gyms and we're getting our own spaces like this. ... Everyone just

> comes because it's free, you come, you play, you talk, you just show one another moves, you say 'you call [that move] this and I call it that'.

Many schools and studios now run them and they are getting bigger, for example, The London Social Pole Club held a pole jam in November 2008 with eight static and spinning poles. Pole jams are, in the main, for polers of an advanced or intermediate level, rather than for beginners, and are a good example of how polers have sought ways to innovate and evolve pole practices and knowledge through community building and social, supportive events (see the photo of a pole jam at the Purity pole studio in the UK in Chapter 7).

Celebrities

The pole community has its celebrities, for example, Elena Gibson and KT Coates in the UK; Jeneyne Butterfly, Mary Ellyn Weissman,[1] Sheila Kelley,[2] Pantera Blacksmith and Fawnia Mondey in the US; and Felix Cane, Suzie Q and Jamillia Deville[3] in Australia. Celebrity polers often travel around the world, hosted by pole studios, and hold workshops

Figure 8.1 Pantera performing the 'Death Lay' after the Miss Pole Dance UK 2008 Finals. Photography: JasonParlour.com

and master classes, usually for advanced-level polers. Jamilla visited pole studios in Las Vegas in April 2009 and in the UK in October 2009; in July 2009 Panetera was on tour in Texas; and Felix, the winner of World Pole Dance 2009, announced workshops in the UK in August 2009. Suzie Q regularly visits Europe to hold master classes and to perform at clubs, mostly recently in New Zealand and the UK in summer 2009; as in summer 2007 KT Coates toured the US. These tours are advertised via Facebook and other forums and are immensely popular, often booked to capacity within days. Many pole celebrities produce instructional or performance DVDs. But the celebrities are not the only faces of pole. Many pole studio websites show a great deal of press releases, from newspaper articles to radio interviews, as instructors continuously attempt to change the perception of pole to the general public; nominating themselves as unofficial spokespersons, with many voices and no one particular accepted 'truth'. As I have said, pole remains anarchic, despite its roots.

Merchandise

Many schools, and most of the pole celebrities (see below), produce some kind of merchandise. For example, there are DVDs by Pantera and Jamilla, by KT, by Elena, by Bobbi, by Mary Ellyn, by Sheila Kelley and by Pole People. Many schools, such as The Flying Studio, Pole-da-Cise and Studio Verve produce T-shirts, tote bags and other items. Most schools also offer, via their websites, the X-Pole and some kind of grip product (such as chalk); most 'high church' schools sell stripper shoes; some schools also sell specially made pole crash mats for the base of the pole (these are usually pink). The X-Pole is sold via many school websites as well as its own, and has recently gone on sale in Russia.

In 2008 in the UK a charity calendar was produced by Kat Williamson of the PoleLove studio in Berkshire in the UK where photographs of 30 polers, from studios all over the UK, were featured (29 women, 1 man) over the 12 months, demonstrating various pole moves. The aim of the calendar was to raise money for cancer research and also to 'show what pole dancing is to me and my friends, lots of fun!' The photographs in the calendar are not glamorous: some of the 'models' look slightly pained or strained as they hold their move for the camera, others are laughing gaily, one or two are caught unawares, and we see some cellulite and not much make-up. For each month there are groups of two or three women (or, in one case, one man and two women), wearing gym clothes such as shorts and vests, everyone has bare feet, some have short hair, some long, one woman has bright blue hair, another's is bright red, there are women who appear to be 'older', that is, in their forties, and there are perhaps four women who are, according to current cultural norms, 'overweight'. Noticeably, all 30 are white. This is not the equivalent of the Suicide Girls website which aims for glamorous, run-of-the-mill (by which I mean young, slim, mostly white, often pouting), albeit 'alternative' beauty and nudity. The photographs on the PoleLove calendar were not even as artistic or well shot (nor were they sepia) as those for the famous calendar where the women of the Rylstone Women's Institute (all in their fifties or older) posed nude for charity in 1999.[4] But the photographs show exuberance, they show friendships, and they show people who are doing something they enjoy. They are not about sexualised display but, arguably, more about the display of community and a sort of engaging showing-off: 'look at this, I can do this!' All in all, the calendar wasn't 'sexy' but it did illustrate the physicality demanded by pole, the skill of the women featured and the friendships of those women.

Events

Pole as performance or competition is being 'mainstreamed' away from its association with strip clubs and highlighted as an athletic activity. Three of the teachers I interviewed for stage 1 mentioned the Pole Princess competition and why it was a positive move towards pole being accepted as a mainstream form of exercise and dance:

> I don't go to the competition because it is really only this year ... this last year or two that they have done pole competitions where you don't have to get your kit off. Until the Pole Princess one this year, they have all been big boob competitions, as far as I'm concerned – and it has not been judged by people that know anything about pole dancing. (Rachel)
>
> I hate it when they lump pole dancing in with, like, wet T-shirt competitions as if pole dancing doesn't need, you know, you need to be fit, there is skill involved. That's why I don't like the competitions, until the Pole Princess one, that one shows the, you know, the skill involved, that it is hard work and can look beautiful without being all about stripping. (Shona)

ANNIE: I am going to go to the Pole Princess competition, very excited about that. We don't have anybody in it this year but we will next year [2007], have to see how it goes, didn't really get the, we didn't get organised soon enough.

SH: Is that a good thing then?

ANNIE: Definitely. It will show how the pole can be artistic and, well it's as much, or more, of a sport as bloody um whatsit, volleyball and those things. You can wear more clothes for pole, in fact, it's not all about getting your kit off.

As Feona Attwood and myself noted elsewhere (Holland and Attwood, 2009, p. 170), this type of rebranding, the attempts to find ways to create a new image through 'normalising' or mainstreaming, is not new. In the late 1950s Barthes wrote about strippers at the Moulin Rouge. He describes how

> striptease is a sport: there is a Striptease Club, which organizes healthy contests whose winners come out crowned and rewarded with edifying prizes. ... Then, striptease is identified with a career (beginners, semi-professionals, professionals), that is, to the honorable practice of a specialization (strippers are skilled workers). ... Finally, and above all, the competitors are socially situated: one is a salesgirl, another a secretary. ... Striptease here is made to rejoin the world of the public, is made familiar and bourgeois.
>
> (Barthes, 1993, pp. 867)

Stripping (as with the pole competitions) is re-imagined as something that 'someone like you' could do, someone who is good at her job and hopes to progress in it. The sex is taken out and instead is replaced with the girl doing the stripping – 'see, she is an athlete, she is a real person'. In the same way, as the teachers said, the Pole Princess competition was about the woman and her skills on the pole, not about the woman as an objectified body on a sexualised piece of equipment. The Pole Princess competition is still being held (for example, the Northern Pole Princess heat was held in Newcastle in July 2009).

There are now more competitions than during stage 1 of the research and I do not have the space to list them all here; therefore, I will discuss just three of the many competitions. The World Pole Sport Federation hosts an annual Miss Pole Dance World event in which there is no stripping and the competitors must be clothed. In 2010 the first Miss Pole Dance Germany will be held in Hamburg. The focus is on the athletic and artistic side of pole dancing. 'Opening closed minds' is the motto of the annual competition, held every autumn since 2005, which is advertised as being 'strictly a sport and fitness event' (although, perhaps confusingly, there is a glamour category):

> Miss Pole Dance was created to change perception to content of pole dance and showcase the unique fitness opportunity through awe inspiring choreographed pole fitness and stunning feats of strength and agility to allow pole fitness become an accessible and accepted form of fitness and highly technical dance form, open to both amateurs and professional dancers alike.
>
> (www.misspoledance-uk.com, accessed on 10 July 2009)

In 2008 the following Miss Pole Dance competitions were held: Miss Pole Dance UK in London;[5] Miss Pole Dance Australia in Sydney; Miss Pole Dance Argentina in La Diosa; Miss Pole Dance USA in LA; Miss Pole Dance European Championships in Amsterdam; Miss Pole Dance Japan in Tokyo; and World Pole Dance 2009 in Jamaica. The latter was a week-long convention:

> Featuring the World's best pole performers, in a showcase event, demonstrating mesmerizing skill with sheer physical strength and stamina through individually choreographed routines. Inspiring women's fitness and passion for life and taking Pole Fitness into the mainstream fitness and dance culture.
>
> (http://www.worldpoledance.com/, accessed on 17 July 2009)

In fact, the last Miss Pole Dance World event was in Amsterdam in 2005 and resulted in controversy within the pole community. Elena Gibson's performance was in two halves, first as a very controlled and elegant display of ballet on one pole and the other half of the act being a more athletic pole display on a second pole. She came first. Later it was decided that she had stripped, which was against the rules, at the point in the performance where she removed her ballet slippers to move over to the second pole. The runner-up, Reiko Suemune of Japan, was then announced the winner. Not surprisingly, the decision caused a lot of discussion and anger among the pole community. But if the polers are to convince 'outsiders' that pole is a legitimate sport or fitness activity they also have to be very clear about the message;[6] for example, the glamour category which implies that pole is still partially about stripper heels when elsewhere on the same publicity it states it is 'strictly a sport and fitness event'.

The UK Amateur Pole Performer Competition 2009, a new competition, has three heats and a final, with three entry categories: Performer,

Intermediate and Expert/Advanced. The aim of the competition is to give amateur polers a chance to perform; entrants must not have been paid in any capacity such as for teaching or performing pole. The rules state that

> [a]ll performances need to be clean fun, no butt slapping and boob rubbing.
>
> No thongs. You may take off clothing such as costumes to dance however you must have suitable outfits underneath. No nipple tassels unless attached to a bra.
>
> Level 1 Only – Boots Allowed.
>
> Level 2 & 3 – Strictly no Boots.
>
> If the dance is thought to be too suggestive you will be marked down for it.
>
> (http://www.ukamateurpoleperformer.com/CompetitionRules.html, accessed on 17 July 2009)

These rules are arguably less ambiguous and make a clear statement about the aim of the organisers, whose approach may be at odds with more strippery polers. Pole Unity is organised annually in the UK by KT who explained her reasons for initiating it in 2007:

> Well, I think there are lots of different attitudes to pole dancing and that is why I did Pole Unity, because there is somebody who says 'oh no it is about empowerment, and no men should do it'. And then there is me that thinks that if we are going to be respected, I think men and women should be doing it. Everybody should be doing it! And then there are other people who believe that it is more erotic and sensual. So that is why I did Pole Unity, so that we could go 'right, all different forms of pole dancing comes together for a show'. And I think the fitness one, because the fitness industry is so massive ... millions and millions of pounds a year are spent on fitness, I think that's the one that will be dominant out of it.

Many of the events are not attended by some instructors who disagree with the approach of the organisers, or who have had disputes with them in the past. Schools and studios also hold regular performance events of their own, such as cabarets, parties and end-of-term shows, to showcase their skills, often in aid of a charity. Here are four recent random examples (of the many available) of charity events being used to raise the profile of pole as fitness while also raising money for charity:

In February 2009 a sponsored pole fitness event was initiated by two 20-year Exeter women, both sufferers of cystic fibrosis who met through pole-dancing. (http://www.thisisplymouth.co.uk/news/Exeter-pole-dancers-aid-charity/article-731845-detail/article.html, accessed on 22 July 2009)

> The Spin City Charity Showcase, Bristol, in March 2009. Proceeds in aid of Breast Cancer Care.
> (http://www.spincitypolefitness.com/page27.htm, accessed on 22 July 2009)

In Kilmarnock in Scotland in May 2009:

> The pole dancing competition and ladies' night at Kilmarnock Fitness Centre was a great success.And the main beneficiaries were 'Break The Silence', a charity which helps and supports survivors of child sexual abuse. The competition on Saturday night was the second held at the centre, which runs various classes in the fun and energetic way to get fit.
> (http://www.kilmarnockstandard.co.uk/lifestyle/lifestyle-news-kilmarnock/2009/05/08/kilmarnock-fitness-centre-hosts-pole-dancing-charity-night-81430-23554314/, accessed on 22 July 2009)

Also in May 2009:

> Nottingham University and Nottingham Trent University are going to battle it out for the POLE WARS 'We're better than you are!' title, whilst raising money and promoting awareness for Breast Cancer Research!
> (from http://www.polewars.co.uk/, accessed on 22 July 2009)

Another example is that in the UK[7] on 1 May every year a World Record attempt is made for the highest number of polers to be performing at the same time; in 2009 (its third year) Equity announced that the 1 May 2009 would be 'UK Pole Dance Day'.

International Pole Federation

The International Pole Federation (IPF) was an innovation launched in 2006, designed to govern and run pole matters internationally, with local representatives who were experienced pole instructors and performers. KT and Elena were representatives for the UK but later stepped down (Alison was a board member); Jamilla in Australia; and Fawnia and Mary Ellyn in the US. It aimed to connect polers worldwide, to offer advice on 'quality assurance' in all matters pole-related, to support pole in all its forms and to endeavour to mainstream pole's image as a legitimate dance or exercise form. Its aims were sound but somehow the IPF failed. As I argued in Chapter 4, the do-it-yourself nature of the pole community has some similarities to the brief but very visible riot grrrl movement of the 1990s: it is all women, it is for women, it is about women increasing their creativity, visibility and production, and it was, on the whole, without leaders. One instructor said:

> It was never going to work, you can't just have a few people who say, one day wake up and say 'ok, now I will put myself in charge', there is no one in charge, there never has been, and everyone might want to be in charge but everyone will stop everyone else from running it.

Another instructor said:

> I went [to the launch event] but it was chaos, you see, it's like this, everyone runs their own thing, it has to be done by everyone, we don't actually even want it to be all under one roof sort of thing.

And another said:

> Unfortunately it cannot be managed like that, it's a very large group of women who are all very different from each other and have a different idea about pole dancing. You can never put it in a little box and manage it. It's all too different.

The IPF still appears to be occasionally active in the US and Australia if the activity on the respective websites is an accurate record. But, ultimately, it would seem that the IPF is not the driving force it hoped to be internationally simply because, as I pointed out in Chapter 4 and above, pole seems to be somewhat anarchic in its approach; there is no one 'leader' or one 'right way'; every single pole performance is as individual as the poler who performs it.

Publications

In 2006/7 a Dutch magazine called *Art of Pole Dance* was launched but was short-lived. There are now two main international pole magazines, both relatively new.

Pole2Pole is a monthly magazine that was begun in the UK in mid-2007. In the December 2008 issue there is the Pole Dancer of the Month (Gidget, also a participant of this research) who has a feature interview, the cover photograph and the centre pages. There is also pole news, pole pictures, a beginners' move explained, an article by KT about pole dancing when pregnant (as she was at the time), some reviews of pole DVDs, reviews of schools, of Miss Pole Dance 2008 and an interview with Pantera; and lots of advertising. The magazine aims to reach pole dancers, exotic dancers, podium dancers, pole instructors and pole students and so occasionally has a slightly schizophrenic feel although it does underline the shifting subject positions I noted in Chapter 4: everyone is 'for' pole even if they are against other types of approach. Here, again, we see the mixed message that pole still exudes because *Pole2Pole* looks like a glamour magazine for men, maybe a soft-porn magazine such as *Nuts*: for example, the cover of the July 2009 edition is typical in that it has a woman wearing lingerie and high heels. It does not look like a magazine about dance and fitness.

Pole Dance International magazine, which is a quarterly online publication from North America, was launched in April 2009 and includes many of the same type of articles: reviews of schools and events, interviews with pole 'stars' and so on. In contrast though, their cover looks like a magazine for women. The cover of the first issue shows a group of 5 (white) women dressed casually in jeans, in what appears to be a pole studio. *PDI* magazine states it is not linked to the strip club industry and aims to 'celebrate the beauty of pole dance advocates in every form – from housewife to pole fitness champion – at every end of the globe!' (http://www.formatpixel.com/project/10941, accessed on 11 July 2009).

It remains to be seen whether these magazines become properly established and whether others are launched which replace them or even succeed.

Equity membership

The Equity Pole Dancers' Working Party was organised by Genevieve Moody and Dana Mayer and meets monthly in London. Its remit is to help provide better conditions for pole dance performers as well as striving to change the public perception of pole, to show that it is not necessarily part of the sex industry. Membership of Equity is, obviously, relevant only for instructors who perform regularly enough to need representation. One of the group's aims was to draft standard contracts for pole dance performances in the media or live, and another for teaching. Equity offers, for example, advice about contracts and tax and insurance for teaching and performing. A recent email informed those on the email list about future plans:

> Do you want to be classified as a sex worker simply because you pole dance regularly? Do you want to have your advertising using pole dance images banned because you are classified as a sex worker? This is just one of the real issues that Equity is helping us with right now.
>
> (31 March 2009)

The Working Party also aims to form and have approved a Pole Dance Committee which would then have more power to protect and advise its members. Working with Equity is just another way that pole is now evolving from its roots in lap-dancing clubs.

This, a relatively short, chapter focused on the international pole community although I doubt I have managed to adequately capture the diversity, creativity and energy of the polers and could give only a flavour or overview of all the work that polers do every day. The classes are a key part of the pole community but by no means the only sociable event which happens, and thousands of women are engaged in perhaps hundreds of activities, all of them occupying the same subject position of 'for' pole, and therefore working constantly to develop and enlarge the pole community. By contrast, the next chapter, a case study, shifts focus from the macro to the micro and examines just one pole studio in the UK and the three women who work there, discussing issues about friendship, innovation and weight. Although most polers participate in the online community to some extent, pole owes its burgeoning popularity and positive reputation to the daily activities of polers 'on the ground', so it is to some of them that we now turn.

9

Case Study 1: 'Empowering Women with Confidence'

> *We make a point of saying 'this will empower you with confidence'.*
>
> (*Evie*)

In these final two empirical chapters I would like to focus more specifically on what I am calling two case studies. I am thinking of these chapters as 'case studies' in that they are an examination of the micro; of stories which have, over time, reflected larger shifts in the story of pole. The previous four chapters dealt with the experiences of the classes overall and with the wider pole community, that is, they were drawn with a rather broader brush whereas this chapter and the next are more intimate thumbnail sketches. The first of these miniatures is a pole studio within a beauty salon in a particular town in the south-west of England where the motto is 'empowering women with confidence' and which returns to some of the concerns of the previous chapters; those of friendships, charity work, weight and fitness and women-only environments. I experienced some dilemmas about writing this chapter and focusing particularly on these three particular women. Obviously, they want publicity for their classes but equally importantly, and obviously, I don't want to be seen to be singling them out in a negative way. As I discussed in Chapter 1, there were several participants who requested that I use their real names rather than pseudonyms. In an attempt to get around this ethical conundrum I will use the real name of the pole studio, which is Pole-da-Cise, although a different name for the gym and salon which it is in – for the purposes of this chapter I am saying, in an effort to maximise confidentiality, that the salon and gym are called The Beauty Spot (although its real name is nothing like that). Additionally, the names of the three women are pseudonyms: Keira, Libby and Evie.

If someone already knows them they will work out who they are; this is always the case and there is always this possibility. If someone doesn't know them this gives them some anonymity. I suggested this compromise to the three women when I visited and it was acceptable to them.

The train to the town takes about 40 minutes stopping twice at two affluent Georgian country towns. The day I travelled there (in 2008) I was aptly re-reading a Jane Austen novel; the countryside was brilliantly green and verdant, just about as English and lush summertime a sight as you could hope to see from the window of a train. The town itself is ancient with a population of about 30,000. I first became aware of Pole-da-Cise because I found Libby's pole dancing videos on YouTube which charted her journey from overweight and, it is fair to say, slightly depressed and very bored young mother to a more lithe, certainly more confident and fulfilled, YouTube personality, Pole-da-Cise marketeer and spokesperson and, latterly, an instructor, photographer and webmistress. YouTube, as I discussed in the previous chapter, is a key way that polers show off their skills and keep in touch with polers internationally, commenting on each other's progress and prowess. Later in this chapter I will examine in more detail Libby's own videos and their progression as she shed three stones in weight.

Elsewhere (Holland, 2009) I have focused on three vignettes of gendered leisure, using the stories of three particular women to explore how they juggled finding the time, the money and the motivation to attend pole classes. The women (Sharon, Parveen and Amanda) were chosen because they had the strongest narratives of 'obstacles to overcome' and ultimately the very fact of arriving at the class was as much a pleasure and achievement to them as actually being at the class. The case of Keira, Libby and Evie is somewhat different since they all spend every day in the pole studio (Keira less than the other two, since she was mostly in the front of the building running the beauty salon); they have no problems with how to access poles, the teaching of pole skills or other women who pole. What emerged strongly from their interviews are narratives of weight gain, weight loss and shifts in body image. Their stories around pole classes are, in the main, about how pole improved their lives, through friendship and through improved fitness and overall confidence, and how they each strive to find ways to enrich other women's lives in the same ways.

Libby, Keira and Evie

I interviewed Keira in the beauty salon, with Libby present. I interviewed Libby at her home on our own. That night I spoke to Evie at length in the pole studio and interviewed her the next day in Libby's house. Libby was 30 at the time of the interview. As a child and a teen she was active:

> [Q]uite a bit of a tomboy and did things that girls don't normally do like rock climbing and stuff like that, archery. ... My boyfriend at the time used to jet ski, so I used to go off and do stuff with him. Because I was small enough and confident enough to do it.

But in her twenties she did little or sporadic exercise until after her children were born, which I discuss further below. She had met her husband at work and had two small children by the time she was 26. Libby was the person I initially contacted because of her YouTube videos, which I discuss further below.

Keira was 39, married and had a teenage son. Her mother and aunt were businesswomen and Keira had taken over the management of The Beauty Spot the previous year. She had a history of sporadic exercise, particularly going to the gym. She is Libby's brother's wife; Libby got to know her better, as a friend, because Evie introduced them.

Evie was also 39 at the time of the interview. She and Keira had known each other all their lives but she was the one who brought the three of them together and initiated them into thinking about pole classes and how to develop them. She was married, had two children and had always exercised:

> My story is from sort of quite a young age I did various forms of dance like ballet and tap, from a young age. And then as I approached my teenage years, sort of early 20s, I became part of a dance troupe and I danced in clubs, clothed, but more sort of podium type dancing – set, sort of, type routines to music that was around at the time. So we are looking at the 80s, so we are going back a bit. So I was always keen on dance the majority of my 20 somethings, if you like, and then I was always keen on fitness so I did things like aerobics and keep fit, general type stuff, like most people. I never had a specific qualification, it was just something that I enjoyed doing and you know, obviously I had a bit of a talent for it if you like, so I continued on with it.

The three women saw, or spoke to each other, every day. Evie and Libby also lived near each other. There was a fourth particularly close friend, whose pseudonym here is Jeanette, and who later became an instructor, who I did not interview or meet. These four women were part of a larger group who regularly met and socialised together in the town.

Pole-da-Cise

Opened originally by Keira's mother and aunt, both businesswomen, The Beauty Spot, a women-only beauty salon and gym, had been open six years when I visited and was run by Keira since 2006. It was she who invited Evie to start pole classes

> not as a pole dancer, but as fitness dance exercise, incorporating the pole movements with fitness ... I didn't think it was that popular at the time – I think I just saw the potential in it. And together we kind of chose the name, Pole-da-Cise and it has just kind of gone from strength to strength.
>
> (Keira)

Of all the studios I visited (ten in all) Pole-da-Cise ran the most classes, all women-only. Their classes ran Sundays 10 a.m. to 8.30 p.m.; Monday to Friday 9.30 a.m. to 8.30 p.m.; and Saturdays 10 a.m. to 1 p.m.[1] The studio was built in winter 2006/7 by 'knocking down a few walls' (Keira) as the pole classes were becoming so popular that it was no longer feasible to run them in the salon. Although this attracted lots of new members it also drove some away:

> KEIRA: A few of my old gym members didn't like it and I have lost some older members.
>
> SH: And why was that? What did they say?
>
> KEIRA: The stigma attached to it – I didn't think people were quite open ... you have got to be open minded about it and actually really take the time to look in on it and watch maybe and sort of like get over the ignorance that maybe it is all about lap dancing and doing things for men's pleasure, do you know what I mean?

Stigma, and the perceptions of others, was mentioned in Chapters 2 and 7, and I return to it in the final chapter. Attempts were made to reassure people that this wasn't about strippery pole or training for lap dancers. Their website states that 'this is a fitness class so heels are not required' and, unusually, that prospective students can go along to observe a class before they join. Classes are kept small with never more than five in a class (there are five poles, all spinning rather than static) and no one sharing a pole. Evie explains her decision:

> We decided that ... for a fitness class and an exercise class, the idea of queuing and waiting and standing around – it no longer becomes an exercise class, because you know, how long are you waiting around or are you going to cool down, could you get cold? Then you could attempt to take your turn and perhaps then end up doing yourself an injury. So I kind of was never keen on that, and I felt that small intimate classes are more likely to ... confidence levels are more likely to come about quicker than if they were with a larger class of people who were having to stand around and wait for their turn, as it were. The other thing is, that within Pole-da-Cise – we work beginners, intermediate and advanced – all work to short choreographed routines. As you get more experienced, then those routines become obviously a lot longer with a lot more moves. So by keeping a class working together, their routines as they progress become sharper. And so when they are eventually more advanced, they are in touch with each other and they know when to slow down, when to speed up slightly, when to catch up with the rest of the class.

For example, in the first class I observed, there were three students who each had a pole to herself: two had been going for eight months and were friends. One of them was very overweight. The third student had been going only a few weeks but had made rapid progress; she was very thin and silent and concentrated hard. Two girls arrived for the next class, both about 20 years old, one was a new mum and the other was a car mechanic. The latter was very strong but the former was determined to get it right. Evie knew their names and what progress they had made; she also remembered things about their lives: she asked the young mum how the new baby was, she asked the larger girl in the first class about her diet and said to be careful to avoid faddy diets.

> You can't, exactly, if we were to line up all our members, then I would probably say three quarters of them you would say 'right, I wouldn't have thought that you would be – this would appeal to you' ... [but] they are finding out the fitness side of it and the values of that. But there is also a lot of ladies who think 'well actually,

> I think it is quite a good way of feeling good about yourself'. And there comes a point where every woman out there wants to feel a little bit sexy, and as confidence builds then, as I say, shoulders drop and maybe sort of you do feel a little bit sexy, because you are dancing as well as working out around the pole. But you are not doing that for an audience; you are doing that for you, because you feel good.
>
> (Evie)

But, like many of the pole instructors I spoke to, Evie and Keira have a philosophy about pole and how it can help women who attend their classes:

> We wanted women to come and join our parties, of any age, shape or size, and to experience the sort of adrenalin rush, as well as the health factors with it and the fact that you can really lose weight, tone, strengthen and just have that real wellbeing. As well as that, there comes a point where every woman ... wants to feel that they are happy with their body – whether they are ... whatever their age or shape or size, and why not? And so what we found when we started Pole-da-Cise, by addressing those sort of issues and offering classes, small classes so that these ladies don't have to feel intimidated. And obviously if we have people ... beginners classes start up, we kind of keep them small. These ladies can come along on their own, and they soon make friends – because over a period of time they are seeing these people quite regularly.

Evie described how students hunched their shoulders at first when they were beginners and how, to her, the sign of them becoming more confident was when their 'shoulders dropped', when they became more relaxed and more able to look at themselves in the mirrors:

> And they are all a little bit nail biting and a little bit anxious and perhaps feel a little bit silly that they are walking around a pole. But gradually as their strength builds and they are able to start doing certain moves, you can suddenly get them glancing in the mirror and you can see their shoulders drop. And yes, they do look good. And then if they look good and they see that, they feel good inside as well as outside. So it is kind of like ... it is almost like having the power to make someone feel good about themselves and walk out of that door thinking 'I feel really great tonight, I have just accomplished

> something – whether I am 18 or 60, that perhaps I would never have thought of doing before – and now I have done it and I am proud'.
> (Evie)

Now Libby also teaches at Pole-da-Cise, although she didn't when I was there. She describes two kinds of student who need to be inspired and made to feel comfortable: the 'wall flower' who is shy and doesn't want to try new moves or variations on moves; and the 'giver upper' who does not have a positive attitude. To motivate them Libby uses 'do what's right for you' which I discussed in Chapters 1 and 3, which she said is commonly used in their classes. A statement on their website is actually taken from the transcript of my interview with Evie and outlines their approach:

> At [The Beauty Spot] our motto has always been empowering women with confidence. Whether it is with the gym, beauty, hair or in the workout classes like Pole-da-Cise. It's all about making women feel good about themselves. For me personally and many of the members, Pole-da-Cise has really empowered women with confidence & inner and outer strength. Whatever your age, shape or size – if you can be empowered doing something that you never thought or dreamed you would that's just fantastic. Try new things that make you feel great is the best way to keep healthy inside and out. To all those people that believe pole work is about the sexual use of a woman's body to attract men, I would say come and see for yourself the fitness levels, inner confidence beauty and strength of our members in the ladies ONLY [their emphasis] environment that is relaxed, friendly and filled with fun.
> (http://pole-da-cise.com, accessed on 2 January 2009)

To Evie, then, empowerment is about individual women feeling good about themselves, about their bodies, and increasing their confidence; the common understanding of the word, as we saw in Chapter 6. Pole-da-Cise was the only school (of the ones I visited, although there are several in the US which do so) which explicitly used the term empowerment as part of their marketing and teaching; their stated aim was to empower the women in their classes.

Friendships

In Chapters 7 and 8 I discussed how important friendships are within pole classes and the wider pole community, and how, as both Alison and Suzie Q commented, there is something 'going on' with pole classes in that women make good friends there. While pole classes had provided the impetus, and the cement, for their friendships, Keira, Evie and Libby also frequently socialised outside of class:

> We go out and socialise with friends and we like to get out and have a good drink ... go away with the girls and have girls' weekends away.
>
> (Keira)

All three women worked hard but also, because of supportive partners, were able to regularly go out, or go away. Libby also found that her initial envy of her friends' glamour, indeed of their friendships, prompted her to make some changes in herself:

> I met Evie, because my sister-in-law married her brother and Evie is a good friend. And you know, going out with Evie and her other friend together when they went out, they looked like film stars – you know, they looked like fantastic. And although Keira's not the same size as them, she kind of fitted in and it all looked ... you would go out and there would be them 2 arm in arm down the road and I would sort of walk behind thinking to myself 'I am younger than you and I should really look similar to you', you know, maybe not as much – I wouldn't go blinging up or tan myself or whatever – but I just didn't feel like I should feel. I was a mum of two at the age of 25/26, you know, and it was like 'oh I should be a bit younger and go out and look a bit better than I do'. So I was a bit envious, but this is before I knew what Evie was like and the type of person she was. I didn't really know her very well. ... She was the first the thin Barbie princess that I actually liked to put it bluntly because I was depressed at being fat and like most women they don't like to hang around with thinnys, especially if they know they are the bees knees. But Lisa was different. I looked past the boobs and blonde haired Barbie and she looked past the fat mumsy me.
>
> (Libby)

As Sarah Whitehead and Stuart Biddle (2008, p. 248) found with their sample of older adolescent girls, they felt uncomfortable and demotivated around other girls who they felt were thinner and/or more beautiful than they were. Ultimately, we see that meeting Evie and Keira and becoming involved with Pole-da-Cise did empower Libby to change her life: she lost weight, she exercised regularly, she started a college course, she began to dress up and enjoy herself when she went out and she felt the support of strong female friendships which sustained her. She also established her own identity as part of the group: she was the 'technical one': the photographer, the one who makes videos and designed the website; although, later, she also became an instructor. Before that, it was being overweight and often lonely which had made her depressed.

Weight and fitness

Weight was mentioned by several instructors, for example, Rachel said:

> You would think it would be the light strong ones who would always be the best, but it is not like that because I get quite a lot of larger ladies who come, and obviously they are holding up more weight – but at the same time, if you are lifting more weight, you are exercising more, you are getting fitter quicker. And you quite often get quite well built ladies who are stronger than really thin people who aren't used to holding themselves up anyway, and they are pretty much in the same boat. ... There are some moves, like an upside down V, because of their body shape, it is challenging not to over-balance ... I give them slightly different things and also, like, no offence to them, but they have got a tougher job because they have more to lift.
>
> (Rachel)

As Frida Kerner Furman (1997, p. 68) notes, 'being overweight represents the clearest failure in the maintenance of an ideal femininity, that is, of a femininity defined by the dominant culture'. Both Evie and Libby said that they felt that they had become overweight after the birth of their children and that this had depressed or alarmed them. Evie described how she had been introduced to pole after the birth of her first child:

> I had a very good friend who was a professional pole dancer, and who was a very sexy lady and just lovely – a very good friend. And we would spend a lot of time with each other, and she had gone on and had children and she wasn't dancing as much, but she had her own pole at home – and so we would often, you know, spend time together and she would work out on the pole in front of me and show me different things. And I found it fantastic to watch. So from there she said 'come on, I will show you some moves'. And when I first tried it ... I had had my first child who was huge, a 10lb 2oz baby at 63cm long and I am not big, you know, not a big person. So I was

> huge, and for the first time in my life I had experienced having to sort of think about my weight. And she said 'well come on, we will do some workouts together'.

Evie and her friend were constrained by their other responsibilities (both had children, partners and jobs) so they practiced together while also forging a strong friendship:

> And so because we had the commitments of children, her and myself, it wasn't like you could get off to the gym and you could go dancing in a club and burn those calories away – so we sort of found time, in between motherhood, for her to sort of show me different workouts and different things that you could do on the pole. And from that I just got a real buzz, an adrenaline buzz, and you know, it got to the stage where we would see each other fairly regularly and we would spend maybe 3 or 4 times a week, spending some time out with her showing me stuff on the pole. Until eventually, I started noticing that I was sort of losing a bit of weight in the places I wanted to lose it, and I started feeling a little bit better about myself as well. And I was able to start doing sort of some pretty good stuff and I felt quite proud. And from there I kind of just loved it, it was a real adrenaline buzz.
>
> (Evie)

Similarly, Libby's weight after giving birth prompted her to begin to exercise regularly:

> I had my first child and went up to size 18, and thought 'well, my priority is my child, it doesn't matter how big I am; that's what's important'. I got bigger when I had her because I was sitting around, on the Internet or telly, and then had [son]. I shot up from a size 18 to a size 22, although my jeans were size 24. So I was quite big ... I saw pictures of myself and thought 'oh! That's bad'. So I started to go to the gym.

At this point Libby went to the gym with a neighbour and did step classes, or other fitness classes: 'within a year I went from size 22 to size 18 again. So I was like "yeah!" But I was still big'. Sparkes (1995) describes 'critical incidents'; that is, events which prompt action. The experience of both Libby and Evie was echoed by other participants who said that they felt they had put on weight or felt underconfident, prompting them to find something that they felt they could achieve and which might be fun, especially pertinent for those who had previously disliked exercise. Whitehead and Biddle (2008, p. 249) report that being overweight can be both a motivator to start exercise, but also a barrier. Libby went to Evie's pole classes with their other friend Jeanette and, at first, neither of them felt confident enough to even try to freestyle the moves they had just been taught:

> [I went with] another friend – and we came along [to pole classes] and sort of had a go. And I'll tell you what, when we started we wouldn't do anything by ourselves. She said 'freestyle', because this is how they used to teach; you used to learn a few moves and then she used to put a bit of music on and then she said 'right then, you are going to freestyle now'. So me and Jeanette would sort of stand there like statues ... 'you go first', 'no, you go first'. And we were nervous. ... For me it was my weight, because I physically couldn't climb, I physically couldn't lift my own body weight off. ... So I just thought 'no', I took it as a joke, this is just a bit of a laugh.
>
> (Libby)

As we saw in Chapter 7, humour can be a way to diffuse tension in an exercise class and Libby chose to use humour to deal with her own embarrassment and nervousness. 'I wonder if women's preoccupation with weight tells us something ... about the appropriate use of space culturally assigned to women. Women move in space in ways that are different from – far more constrained than – men' (Furman, 1997, p. 71); one might say, as Libby did, that they 'stand like statues'. But, as I argued in Chapter 3, pole teaches women to be less constrained, to extend their arms and legs, to climb the pole, to point their toes, to occupy more space. Pole instructors also have the added zeal, about pole and its potential for women, which benefits the student. As Libby said of Evie:

> And she would [make us do] – like I said last night, one of the easier – not easier moves, but one of the less strength needing moves, so to speak – the frog, what we call the frog is going round backwards on the pole, gripping your legs and turning in. It took us weeks ... weeks and weeks and weeks to do it. And if you see the girls now, it is like they do it in their second [class]. But it took us weeks because we were bigger. ... Her determination to get me to do things even though I was fat was beyond patience.
>
> (Libby)

On Libby's YouTube channel she documents what she calls her 'pole journey', where she films herself as she learns to use the pole, losing weight and building strength and confidence. The videos are literally a diary of her learning to use her body in less constrained ways. Her introduction reads:

> This is my pole diary, my journey! I have an addiction to Pole dance exercise ... I started learning Pole in 2006 and at the beginning of the year not shown in these videos I was a size 22 and just under 16 stone in weight. I don't claim to be a great dancer, I just really enjoy sharing my trial and error moments along with my achievements. If you don't think you can do this and you think you are too large to try, take a look for yourself. It's not all about getting everything perfect and it's not about being tiny, it's about building strength and having fun. After 2 years I now teach at Pole-da-Cise along with my dear friends.
>
> (http://www.youtube.com/user/oliviashelster, accessed on 20 June 2009)

Her channel has had (as of July 2009) nearly 30,000 views and she has 728 subscribers. An early video, and one of the most popular, is called 'Never Say Goodbye to My Pole' and shows a montage of her trying out different moves in the Pole-da-Cise studio; the music is Sarah Brightman singing 'Time to Say Goodbye' (also known as 'Con te partirò', an Italian classical song). Libby believes that the popularity of her YouTube video was partly because she was overweight at the time which encouraged other women to feel able to try pole or, at least, inspired them to exercise (although she doesn't look particularly large, just not as tiny as cultural norms demand) and because of the classical music.

LIBBY: That one is the most popular one out of all my videos. But I know why that is.

SH: Why?

LIBBY: That is because it is a different style of music.

SH: Right.

LIBBY: It's weird, because there is hardly any pole dancing ... there wasn't when I started, any pole dancing videos on there that had anything other than R&B and fast music. Yes, which I got bored of, and I thought 'oh this is a bit like ballet'.

SH: And it did – it has got lots of comments on it, hasn't it?

LIBBY: Yes, it has got the most comments, that one of mine at the moment. And most of them are from people that want to do it but they didn't think they could, which is really nice.

SH: So will you keep doing that?

LIBBY: Yes, I have put some funny ones on there now as well, because I think a lot of people don't have their outtakes on.

SH: No.

LIBBY: I know, they cheat – and I thought well I am cheating a little bit as well, because some of the women even commented and said 'oh you are so good – I could never do that', and that is the same sort of comment as I would put on these previous women's ... smaller women's videos. And now I am getting slightly smaller than what I was when I started ... I am actually smaller now than some of the women that start, that want to start and they don't think that they can. So I thought well, rather than showing all my progress, I will show my faults as well and get people to realise that.

The videos of outtakes, for example, one is called 'Trial and Error', also show that in the most recent ones Libby is now incredibly strong. The comments on Libby's videos usually praise her and are often from women who say they are overweight and she has inspired them to try. The archive of videos means that a viewer can literally watch, in sequence, how Libby's pole skills accrue, her strength and confidence increase and how her weight decreases:

> I was 16 stone 5 when I started and now I'm 13 stone 11 so not a huge weight change but I was size 22 when I began and size 14/16 now ... I don't mention weight to many members but I do mention inches.

Perhaps because Evie and Libby have had their own weight problems, Pole-da-Cise began classes for 'larger ladies', what they called their FX classes.[2] Evie discussed with me the first 2 FX students:

> Yes, we have got 2 ladies who are of the larger frame, let's say – and they came ... they had heard about Pole-da-Cise, and the first thing they said was 'we have heard it is a really good way of getting in shape and perhaps burning some calories and a cardio-resistant workout – do you think we would be able to do it?' And at Pole-da-Cise we never turn anyone away! They said 'we hate the gym' ... and we tried to explain, 'well, we have got ladies of all different shapes

> and sizes here'. But they have obviously been and it has put them off, it has put them off for life because they had gone to the gym, they had been surrounded by lots of ... [thin women] – and maybe they had been to gyms where there were gentlemen as well. And they have just felt intimidated and they have not enjoyed the experience at all – and I think it has knocked their confidence.

Evie was at pains to distinguish the appearance, approach and experience of Pole-da-Cise from other gyms where women might feel intimidated and demoralised while exercising next to very fit slim women and men. As Maxine Leeds Craig and Rita Liberti (2007, p. 682) found:

> The word 'comfortable' arose repeatedly in our interviews. ... How was their sense of comfort produced? Did it arise spontaneously in the absence of men? Is it present wherever a number of women are together? We found that the comfort was provided by an organizational culture of nonjudgmental and noncompetitive sociability and that the foundation of that culture was the organization's use of technology and labor. The equipment and physical setting, the established procedures for customers' use of machines, and the interactional styles of employees fostered a feminized culture in which women avoided criticizing other members, did not try to distinguish themselves from others through demonstrations of greater physical fitness, and participated in conversations while working out.

While Evie was saying that she felt Pole-da-Cise would be a more comfortable and comforting experience for the FX ladies (a feminised culture, a pleasant environment), her concern was also about increasing their confidence – in fact, empowering them with confidence:

> They had heard that Pole-da-Cise is a bright studio, maximum 4 to a class, so they are getting pretty much personal one to one. And I think they felt that 'maybe we could try this'. And after inquiring, 'could we do it, because we are bigger ladies?', and we said 'yes, definitely – it doesn't matter how long it takes us to get there, but yes, we could – but while we gradually work ... I mean some classes might move quicker than others'. Anyway they came along and they gradually built up a bit of strength there, and we worked on them so they were getting more grace around the pole – and albeit they weren't moving on at say the normal pace, they were still getting something out of it and we still felt that. ... They were gradually getting their

> confidence built up. And on numerous times in the early stages, they would ... they would just stop and one person in particular would have a look of a bit of panic, you know 'I just feel that – I am really enjoying it, but I just feel like am I looking silly'. And you know, it was a case of not like totally bigging this person up, but saying 'what you are doing is, your heart rate is getting up, you are moving nicely round the pole, you are getting your legs moving. ...' And you know, as I say, these are bigger ladies – so just a form of exercise with getting them to bend and stretch and do things that they wouldn't necessarily do unless they were going to the gym.

Obviously the process was a gradual one; Libby described how she and Jeanette would 'stand there like statues' out of sheer panic, and Evie said the FX students also would sometimes have a look of panic. The mirrors also seem to be an issue initially:

> I think the problem they would have, or the problem they had in the beginning was within the studio they are facing a huge mirror – and to begin with their problem would be ... you could see that they just did not want to look at themselves in the mirror – and it was confidence. However as time has progressed, we have suddenly ... shoulders have started to drop, and they are now, you know, instead of coming into class ... they are coming into class with a smile and they are happy, and for the first time they were able to attempt, and do, and do very well, and actual spin around the pole. And it wasn't a basic spin – it involved precision, if you like, and they moved into this spin after a short routine. And, you know, they just burst into tears, and it was quite overwhelming. ... So they walk out empowered with confidence, and that is what it is all about, to make someone feel that good.

The experience of being 'shy' of the mirrors echoes Price and Pettijohn's (2006, p. 992) findings that 'body image concerns in dancers may be partially explained by the presence and use of mirrors in the dance environment'. But, as Evie said, the students did feel able to look in the mirror after some time, especially after they had achieved a spin. In fact, they 'walk out empowered with confidence'.

Empowerment

Collectivity, as we have seen, is a key aspect of the pole community. Accordingly, the success of Pole-da-Cise and the increased confidence of

Keira, Evie and Libby are due, in great part, to their strong friendship, their entrepreneurial approach, their concern to provide 'empowerment' to a diverse range of women through pole classes, their energy in raising money for charities and the support of their partners and families. For example, Keira was very clear that without the creative input and energy of Evie and Libby she wouldn't have thought that her business would be as successful, nor that the three of them would be so close:

> I think what has happened is, it is kind of like ... it is all a bit mad. I am not a motivated person, I need to be motivated. And apart from the gym ... I mean I can motivate people in the gym, but with other things I am kind of quite laid back and I am a bit like my mother, 'if it's not broke, don't fix it'. Although I did sort of like do this and that. And then Evie is a bit of a driver, isn't she Lib? She is a bit of go-getter. And with Evie and Libby, they go mad: oh my God, they have got an idea! [laughter] And it is all sort of like – it is just coming all together, you know it is all coming together now.
>
> (Keira)

The ideas that Keira mentioned included entering The Beauty Spot or Pole-da-Cise into competitions for best business, or gym or hairdresser, in the town or country. The results have been a variety of nominations and awards, for example, it has won the County's Best Gym 2007 and 2008. When I was there it had just won the UPS Business Award for 2007:

> KEIRA: The UPS Business Award 2007 – yes, absolutely fab. I phoned Libby and I said 'fill it in first, you never know, we might win'. And we did! And it was like what we had to offer ... what our company had to offer above the others. ... Like what other businesses offer.
>
> LIBBY: The service ... the outstanding service.
>
> KEIRA: And that was judged by Theo Paphitis of the Dragon's Den.[3] ... So we got a month's free advertising with [local] Radio for that.
>
> SH: Is it still on that? [to Libby] Is that why you kept switching the radio on today?

Evie mentioned this kind of teamwork as the reason that Pole-da-Cise and The Beauty Spot have been so successful:

> To start ... although I have been teaching for a couple of years, I have only been running my own business since last year, and it has been

> pretty tough but I have had excellent, good people behind me as well – Keira, Libby, Jeanette – and they are all really close good friends and they kind of think the same way as I do. So I would like to think that my hard work and their hard work is going to pay us back in dividends. ... Everyone has got a bit of input and got something to say and everyone is interested. Because no one looks sort of like strippers or whatever; we are all pretty much 30-something women that are ready to listen and think 'oh yes, that's how I feel', you know we have all got something to say about it. And each comment is different and valuable, and so we have used that, and that is one of our keys to our successes.

Pole-da-Cise offers an example of a successful women-run, women-only business; it offers a feminised, supportive environment in which to learn a skill; the instructors and owners strive to 'empower women with confidence', they continually raise funds for charity and their efforts have been acknowledged by various business awards. The basis of this approach is the close, trusting friendship of Libby, Evie and Keira.

The next chapter, the final empirical chapter, examines another case study, that of men and masculinities and pole classes.

10

Case Study 2: Power Moves and Everyday Bodies

> *I am the only man in a class of 8 and no one bats an eyelid. It's great!*
>
> *(133/M29/UK)*

In this, the final empirical chapter and the second case study, we turn our attention to pole classes for men. It is not difficult for boys and men to find positive, active role models such as action movie stars (Tasker, 1993) but particularly sports players (Horne et al., 1999; Armstrong, 2001) including 'alternative' or extreme sports (Borden 2001; Wheaton 2004; Robinson 2008). As Connell (1995, p. 54) argues,

> the institutional organization of sport embeds definite social relations: competition and hierarchy amongst men, exclusion or domination of women. These social relations of gender are both realized and symbolized in the bodily performances. Thus men's greater sporting prowess has become a theme of backlash against feminism. It serves as symbolic proof of men's superiority and right to rule.

Similarly, Vicki Robinson (2008, p. 62) notes that 'sport is crucial in the maintenance and reproduction of a specifically masculine identity'. (Although, of course, not all men like sport, to some an outrageous claim.) Nonetheless, as David Morgan (1993, p. 69) points out, 'if, in general, the sociology of the body is a relatively late arrival on the scene, the sociology of the male body would seem to be even more of a newcomer. … This tendency to see women as being in some way more embodied than men is reflected in popular culture and popular imagination'. The image of men's classes differs contextually: in a patriarchy the impact and associations of men doing pole are entirely different; there is no history of heterosexual men's bodies being objectified and displayed in the same way as women's have been. However, as Sarah Grogan (1999, p. 58) notes, 'over the last decade, psychologists and sociologists have become increasingly interested in men's body satisfaction. This is largely due to the fact that the male body has become more "visible" in popular culture, producing interest in the effects of this increased visibility on men's body satisfaction'. In fact, Michael Atkinson (2008, p. 67) argues that men are now much more likely to use cosmetic surgery in their response to a 'crisis in masculinity' to achieve 'an acceptable corporeal performance' but that, while the literature on women's experiences of cosmetic surgery is 'rather full ... incredibly few body theorists have empirically addressed men's embodied interpretations of the cosmetic surgery process' (ibid., p. 73). Pole reflects this sort of shift. During stage 1 of the research pole for men was not common although there was some provision, for example, during stage 1 of this research the London-based company PoleCatz (later known as Pole Central) ran classes for men under the name PoleDogz, 'run by men for men'; classes were 'professional and fun'. It was during stage 2 of the research that men began to join classes, or enquire about classes, or that classes solely for men were run. But even this was a gradual development as Jess's reply illustrates:

SH: So, what sort of range of people are they? They are all women?
JESS: Yes, they are, but funnily enough I do get quite a few enquiries from men, but they have never gone through with it.

This chapter is drawn from the data of the three questionnaire responses from men:

> 11/M44/UK is a married IT Executive with 1 child; he has been doing pole classes for 2 years. Found classes 'quite costly'. Has not previously exercised and has an OK body image. He particularly enjoyed the sense of achievement. One benefit is that he is now less shy as a result of attending the classes.
>
> 131/M30/UK is co-habiting with no children, and is a developer. Has always exercised and his body image is OK. His partner is a pole instructor.
>
> 133/M29/UK is single with no children; his ethnicity is mixed (the other two respondents are white) and he is training to be a pilot. He has always loved exercise and his body image is 'excellent (I look good)'. He has been doing pole classes for about 6 months, particularly enjoys spins, and has no difficulty paying for his lessons.

This chapter also draws upon an interview with James, a professional ballet dancer and poler, and from observation of James performing with his dance company The Aussie Pole Boys at a burlesque/supper club. James is in his thirties and dance and performance are central to his life:

> I actually teach dance for a living – I teach classical ballet and contemporary dance ... I still have my own business, I do corporate work and have about 3 bookers, so I get probably about 2 jobs a month which is quite high profile big things and corporate events and that kind of thing, plus working with the drag shows – not in drag myself but as a boy dancer, and I have been all over the all round the world doing that with different drag queens. ... I do a lot of choreography – and for the last 4 years I have been working with some of Sydney's best high schools, private ones ... they do big musicals every year.

These four participants provide some contrast to each other, indicating that pole does not necessarily appeal to a particular sort of man. However, what did the women polers think of men doing pole classes?

Who does pole?

There were two prevalent opinions about men who do pole. The first was a suspicion that they were there (in mixed classes) to 'perv', that is to watch the women in a sexualised way.

> If you had [heterosexual] men in there I think other women would think, 'I've come here for a women's class and there's a guy in here, that's not right, it's pervy'. (Tia)
>
> I wouldn't do it if there were men there: if men walked in I would walk out. ... I wouldn't be comfortable, it would change it. (Charlotte)
>
> If there's a straight guy coming in, we will say, 'yes you can join the class but we are going to watch that first class and if you're here to pick up, you're out'. So it's not a pick-up joint but most guys wouldn't dare. (Bobbi)
>
> I prefer now to keep [men] out of my classes because it changes the whole vibe. I am happy to run courses for men because if they genuinely want to pole dance they won't mind if they are with other boys but if they are there just for a perv well then. ... If I have fought my girls for 4 weeks to get them to feel confident in a pair of hot

> pants and a guy walks in, it changes the vibe, it like undoes my hard work. (Suzie Q)
>
> I was able to talk to the guy and he said 'yes, I want to pole dance, can we do a mixed class?' and I was like 'no, because you want to perv the women and use it as an excuse for a dating service' so I was like 'no, no, no'. (Genevieve)

We see that men or boys are not always welcome as other work on mixed gyms or P. E. classes has found (for example, Flintoff and Scraton, 2001); indeed, there is a very strong belief that men would disrupt the 'vibe'; that is, that a man would undermine both the body confidence of pole students (which, as Suzie Q said, is often fought for, and worked towards, over many weeks) as well as the comfort for women of being in a feminised, all-female environment. Not only that but it is feared that men are there to 'perv', to use the classes as a 'pick-up joint' or a 'dating service' – certainly the polers are suspicious of men who wish to pole, possibly because pole is seen to be run by women for women, although traditionally performed by women for men, which is where the suspicion arises from. But there are few men in classes, and originally there were none at all, which demonstrates to us that pole classes are not about performing for men. In fact, they are very much about creating a feminine and feminised, heteronormative experience where women are not subject to the male gaze, however unpredatory that gaze may be. One respondent met his partner at a pole jam:

> In addition, and indirectly however of getting into pole dancing for fitness as such, a friend of a friend I'd met at pole jams, became my girlfriend! [now wife] This was certainly never my intention of starting the classes, but was a very nice bonus as we're so well connected on so many levels.
>
> (131/M30/UK)

Obviously people will almost always meet their partners socially, be that work or leisure spaces, or increasingly, online. But polers need to be wary about men joining classes, even 'high church' classes, because of pole's roots in lap-dancing clubs. If pole is to continue to evolve and become a dance and exercise form in its own right – related to but separate from clubs – it has to ensure that it does not replicate the idea of pole being about women's bodies being available and on display for men.

In stark contrast, the second belief, or experience, about men who wanted to pole was that they were gay or, as Bobbi put it, 'or should be':

> BOBBI: Men? Okay, we don't run them separately; guys can join in with girls.
> SH: They can do a mixed class?
> BOBBI: Any class, if a guy calls up and says he wants to do a class, we tell them they're starting with the girls; we don't make special preferences for boys.
> SH: How often does that happen?
> BOBBI: We've got about 4 or 5 guys now.
> SH: Are they usually gay men or ...?
> BOBBI: Well, gay or should be! That's what we say. Is he gay? No, but he should be. Or he is and he doesn't know it. So it's generally that kind of guy who comes in and they may come in the dressing room and put a skirt on, all sorts in here but I don't judge, I don't care.
> SH: What do they do about the clothes?
> BOBBI: Yes that's what I mean, they wear skirts and hot pants and stuff. Oh yes, I've seen a couple, we've had to pull a couple of guys aside and say 'we don't mind if you wear a skirt but you need to package up underneath a bit more because when you bend over you're upsetting the girls behind, you're upsetting me and I've seen a bit', so we dare to have a little word. I'm happy they've found a place they feel that comfortable, so I am more than happy for guys to come, it makes me so happy when a guy walks in.

In the same way, then, that strippery pole classes offer women a space to 'dress up', they offer some men a similar experience. It should be noted that gay men's bodies have been sexualised and marginalised much as women's have; and that men who do not fit into the narrow masculine template (such as transvestites) will often have to seek out somewhere safe to experiment. Some men, of course, fit into neither category of 'perv' or 'gay, or should be' but face this sort of initial wariness because, as I have argued in previous chapters, pole is run by women for women and men joining classes ruptures the feeling of safety, instead replacing it with feelings of discomfort or embarrassment (James, 2000; Whitehead and Biddle, 2008).

Why do it?

If the presence of men in classes is so potentially disruptive, and if men are, to varying degrees, unwelcome to join all-women classes, why do they attend? As Michael Gard (2001, p. 223) argues, there are a whole range of issues around masculinity and dance, not least the conundrum of how to make dance appeal to boys who are concerned about preserving their own sense of appropriate (or hegemonic) masculinity without alienating children 'who have found physical education all too "boy friendly" in the past'. Arguably, these same issues apply for men. Alison explained that most of the men who had come along to Pole People classes were already dancers:

> The guys that came along were more fitness-y than the women – they weren't all gay guys at all, it was about, um, I'd say about 50/50. A lot of them were fitness instructors, or tai chi, or dancers – so they were more into the fitness and dance industries, more likely to be professionals, than a lot [of women] who come to classes. ... More men came individually [than in pairs or groups] and they were obsessed with power tricks! ... I mean, when the guys do spins, you know, they have so much power, it's brilliant.
>
> (Alison)

James and two of the respondents did have a history of regular exercise but one respondent, the oldest male respondent at 44, did not:

> Prior to 2005 I was not really into fitness, although as a child I quite liked watching athletics and wished I could [do that]. Never mentioned this and didn't have the opportunity. Bit late now for this discipline, but pole would fill the gap well.

The questionnaire asked 'why did you start to do the classes? What attracted you to pole?' to which he replied:

> Exercise with a dance and gymnastic side to it.
>
> (11/M44/UK)

His love of watching athletics as a child and his history of non-exercise dovetail with pole's athleticism and its ethic of 'do what's right for you'.

Another participant also echoed the sentiments that pole is fun and avoids being as 'mindless' as other exercise classes, as participants such as Lilia also mentioned:

> The great fitness aspects for the upper body, for improved strength and flexibility, whilst being a fun environment, a challenge (i.e. aiming for the next tough move not just about x more repetitions), and to the accompaniment of music which always helps good exercise in my experience!
>
> (131/M30/UK)

The third respondent wrote:

> I love being the centre of attention.
>
> (133/M29/UK)

This response implies that the respondent expected to be a novelty in classes as the only man and that he saw pole as a performing art rather than as primarily exercise. The other respondents saw it as both, indicating that the novelty appealed to them but also the idea of exercise which was fun and which allowed for feelings of achievement.

James's background was classical dance but he began pole, via a circuitous route, because of a friend who was a drag artist:

> I went to the National Ballet of Portugal in Lisbon and then I went to the National Ballet of Canada in Toronto and I came to Sydney Dance Company in Sydney ... did a lot of work with the opera and I formed my own business which does dance [and] which I still do. And I also have been working with a drag queen [name] who is actually French, for about 8 years – and the reason I learnt the pole dancing was because of her – I was doing a show with another drag queen and there was a pole and we used to play around with it and do stupid stuff, and then she actually manages the Slide nightclub here. ... And she said Suzy Q and Candice – Candice is Miss Pole Dance Australia at the moment [2008] – are performing, and she came up to me and she said 'oh you can pole dance can't you' and I said 'sort of', and she said 'oh could you learn and I can put you on', so I did 3 lessons [at Bobbi's] and she threw me on to perform and that was about 2 years ago now.

The physical benefits were mentioned by all the male respondents, for example:

> It is so physical – my body has never looked as good as it has since I have been doing pole ever you know, and I have been a professional dancer all my life. It is just so good for strength and just toning up your entire body, it is amazing. ... It is really dangerous stuff, like hanging up by your legs and I was nervous I was going to break my neck, you know, that was why I was nervous [before his first performance], not because of the performance at all because that is the easy bit for me; it was the actual physical aspect of it that is bloody hard.
> (James)

This kind of response echoes those of the female polers in previous chapters (that pole is hard work) and illustrates that pole is not just of benefit to women: men also feel definite physical benefits as well as enjoying the friendships and freedom of the classes.

Where is it?

James taught mixed and women-only pole classes at a dance studio. The three questionnaire respondents had different experiences of the locations of their classes:

> There have been so far a variety of locations, from pubs, gyms, dance studios, out of hours nightclubs, etc. (and of course with my own pole at home or at the instructors home). (131/M30/UK)
>
> A club. (133/M29/UK)
>
> 1 private in [town], 2 when running in [town], 2 pole jams near [town]. Now looking for other options. (11/M44/UK)

Many pole schools offer mixed or male-only classes on their websites, for example:

> Can men join a course? Yes men can join the courses, depending on what you want to gain will depend on whether you can join a beginners course or go straight to intermediates.
>
> If you would like to do the whole range of pole dancing and tricks then the beginners course is essential. Or if you only want to do the hard strength moves then intermediates would be ideal.
>
> Men are also welcome on the party bookings. (http://www.puritypoledancing.com/faq.html, accessed on 11 June 2009)

> Chinese pole classes also available for BOYS – great for upper body strength! (ZebraQueen, July 2009)
>
> An intensive guys only class which aims to get you to a level that will wow your friends on the dance floor or build muscle tone and stamina in a way that other sports simply cannot, Pole work for guys combines a cardiovascular activity with a muscle building work out. Please call us on [number] to book your place in our guys class on a Monday at 5pm or a private lesson at another time. (http://www.poledolls.co.uk/, accessed on 16 June 2009)

Some of the longest running mixed classes are run by the York University Pole Exercise Club (YUPE) in the UK.[1] Despite its name the club's FAQs state that high heels can be worn to class and the website sells a large range of stripper shoes (only in women's sizes); in fact, the YUPE Facebook page states: 'please be responsible when posting images to this site, this is not a pole dancing forum and as such photos must be typical of pole exercise classes. No nudity, shorts and T-shirts only please'. This kind of mixed message, discussed in Chapters 5 and 8, is part of the reason that polers have to be cautious about including men in classes; stripper shoes will increase the possibilities of attracting 'pervs' to the classes because stripper shoes, traditionally, are worn for men to 'perv' at. However, the York classes do include men:

> Can men take part? Yes! In fact the male membership to Pole Exercise Club has risen dramatically in the last academic year. Men begin the exercise with a natural strength advantage, but [name] our male instructor utilises the unique aspects of Pole Exercise to build muscle and strength, increase flexibility and push you to a different level of fitness. This exercise even benefits the athletically involved individuals, such as gymnastics, climbers to even skydivers as it works to specifically enhance the area in which you wish to work on!
>
> (http://www.yupe.co.uk/faq.php)

YUPE's attempts to masculinise pole include mentioning traditionally 'male' activities such as climbing and skydiving, equating pole with a reckless, strong, masculine type of physicality. Kiwi Pole in New Zealand (also mentioned in Chapter 6) runs classes which include pole for men:

> HisPole – These classes are for Men only, sorry ladies. All about building strength, concentrating more on Chinese pole style of pole

> fitness. No dancing involved!!! Any age shape and size welcome! Mixed levels.
>
> (www.KiwiPolefitness.co.nz/, accessed on 9 July 2009)

Interestingly, Kiwi Pole go as far as reassuring men that there is 'no dancing involved' as if men do not dance; they also reassure prospective students by welcoming 'any age shape and size' which is how many schools also market their classes to women.[2] This type of reassurance feeds into the discourse about men and body image, what Michael Atkinson (2008, p. 69) calls a 'crisis' of masculinity:

> In the midst of the perceived crisis, certain men refuse to acknowledge or embrace new masculinities – despite popular discourses regarding metrosexuality or ubersexuality – and retrench into traditional, essentialized and hegemonic masculine images and embodied performances. Yet others, however, are discovering innovative ways to reframe their bodies/selves as socially powerful in 'newly masculine', or even what we may call 'male-feminine', ways.

Atkinson is discussing cosmetic surgery as a way that men choose to find 'newly masculine' embodied practices but, similarly, men attending pole also do just that: while pole is still often perceived as a sexualised activity for women it is also increasingly being seen as a sexy and powerful activity where polers build muscle strength and skill; it is also a heteronormative environment so that men who pole can often say that they spend lots of time with women polers, at classes or pole jams. Pole can become a way to be 'newly masculine'.

How does it differ?

Pole classes, as we see above, are marketed as being hegemonically masculine and about 'power moves', muscle building, strength and stamina. (Similarly, the difference between men's and women's football/soccer is commonly accepted to be that the former is faster and stronger while the latter is slower and more skilled.) Although classes are marketed among both men and women as a tool for building upper body strength and toning, they are also marketed as being inclusive, friendly and empowering among women, which, for men they are not. As Sarah Grogan (1999, p. 69) argues, 'the activity most obviously linked to improvement in body image for men is weight-training and body-building, activities that would be expected to lead to development of muscle mass, to bring the male body more into line with the mesomorphic ideal'. And yet male bodybuilders often receive negative reactions to their bodies which are seen as 'over-done' or 'unnatural' although, arguably, less so than female body builders (see Mansfield and McGinn, 1993; Tate, 2000; Grogan et al., 2004). Men's pole has its 'celebrities' just as women's pole does, for example, AJ in the UK; Timber Brown and Bad Azz in the US; Cuban twins Yoeny and Yodeny; and Dominic Lacasse, a Canadian gymnast. The latter is famed for his feats of strength such as his ability to hold the position of the 'flag' move for 39 seconds (in 2007) for which he holds a Guinness World Record. As we saw in Chapter 3, 'in competitive sports, women's bodies are sexualized but men are portrayed as powerful' (Kotarba and Held, 2006, p. 153). One respondent wrote about pole:

> Males [are] known for power moves, flag hand stands etc; quite a few rugby players, and parkour[3] males have moved over for the challenge.
>
> (11/M44/UK)

James said that male polers were more likely to be able to climb the pole more easily but refuted the idea that strength made pole easier per se:

> Because of my training I've got a certain aesthetic look to what I do and that is why it works. ... I loved it, and because, not that it is easier for guys, but guys have a different strength; guys love upper body strength which makes it easier to climb up the pole but it is just as hard to get in a good position on the pole.

Some instructors discussed how teaching men also differed from teaching women. For example, Genevieve discussed some of her experiences of teaching men and how it differed from women, for example how use of the mirror differed and whether male students listened to her instructions:

> Occasionally I do get guys but I'm thinking of knocking it on the head [stopping it] because it's hard to get them through the door. ... The only reason I offer men's pole classes, I'm not that keen on it, but I'm trying to give it as much of a go as possible, I think 'well, why not, if women can do it there's no reason men can't' but I won't teach them the same way. I am not going to teach men how to pole dance like women. ... When I teach women it helps if I face the other

way and then do the movement and then they can copy me [in the mirror], men haven't got that mirror image issue so I can look at him and show the move and they can get it straight away. ... But the only thing is, they don't listen, so if I'm saying 'show me a spin' or 'put your hand there and the other hand here', they get it wrong, I show them again. It has to be a bit more shouty!

(Genevieve)

Women's use of the mirror is, as I discussed in Chapter 9, often about a lack of confidence and an unwillingness to look at themselves. She also explained a distinction between the content of classes for men or women, and acknowledges that teaching in these different styles would benefit her teaching:

If I do a guy's class it will be purely, there will be dancing there but they won't be doing ... body isolations and stuff [such as hip gyrations], they will be doing it like the difference between male and female ballet dancers. I'll make them climb just with their arms and things like that. I mean, it will be good for me too, for my teaching.

(Genevieve)

James also mentioned the difference between men and women in class and how that impacts on what he teaches:

JAMES: Okay, first of all women's flexibility generally is much easier than men's flexibility; I always start my classes with a full stretch and try and getting the guys down to splits is really hard as you can imagine.

SH: Yes.

JAMES: Also the ethic of women and men. I think for women they need to be able to do the whole body roll – strippers ethic in there even if, it doesn't have to be like a slut, it can be done artistically. For me it doesn't work for men at all unless it is a very specific show for a very specific crowd and just because the physicality of men is so different we push that with the guys much more. One thing that is very similar though I think is the control, you know, if you see a woman being really controlled and everything is excellent and smooth; it is exactly the same for guys even although the physicality is different. ... The boys because of their strength tend to learn more quickly even though it

> may not be done as nicely. ... In my class I get the girls to walk around on demi pointe even if they have got shoes on, where the boys are on flat feet, you know what I mean, there is little things. ... That is just my ethic, and other people are different.

James's comments reveal that his approach to pole is more 'high church' or, at least, more about performance than exercise. It also reveals a division between how he views male and female polers. Men can be smooth and controlled like women can but they look different, so pole itself looks different. Men are on flat feet, not up on their toes or wearing stripper shoes; men are also less likely to do the splits; men do not do body rolls and other linking moves which have a 'stripper ethic'. James states that such moves needn't be done 'like a slut, it can be done artistically' – nonetheless he doesn't teach men to do the same moves because men's embodied practices are different to women's. Bobbi's comment about men 'gyrating on a pole' echoes James's:

> Men doing ballet I love to watch, they're so graceful and beautiful and that's as feminine as you can get and I love that, but pole is not feminine, pole is very sexual and to see a man gyrating on a pole doesn't really sit right with what's in my head.
>
> (Bobbi)

So although male ballet dancers are graceful and beautiful Bobbi cannot imagine men on a pole achieving the same effect because pole, and therefore male polers, are different; yet when I watched James perform on a pole, his performance was graceful, beautiful, strong and sexual. But traditionally, as I discussed in Chapter 2, men's bodies are not objectified or used in sexualised display as women's are, so it is harder to make the connection. Exercise polers have sought to short-circuit that association so, as I described in Chapter 4, women in pole exercise classes (where men are the least likely to be included) wear shorts and T-shirts just as they would to the gym, with bare feet. In strippery classes (where there is more of a chance of a mixed class) women also wear shorts and T-shirts but also wear stripper shoes some or all of the time, shoes which are automatically sexualised in the popular imagination. All the men I have seen doing pole wore shorts and T-shirts and bare feet; as James said, 'flat feet', although they are encouraged to point their toes just as women are. So what do men enjoy about pole if it is different for them? When asked what they particularly enjoy about classes the respondents said:

> Just the moves and tricks and the challenge/achievement. (11/M44/UK)
>
> Spins. That simple. (133/M29/UK)
>
> The diversity of challenges as mentioned above (there's always a new move to work towards to keep focused). Also the collaborative spirit within 'pole community'. (131/M30/UK)

And when asked 'what are the positive things that pole classes have given you?' they replied:

> Less shy. (11/M44/UK)
>
> Improved strength and flexibility on a physical level. A great bunch of really excellent friends is a significant addition elsewhere. (131/M30/UK)
>
> More confidence about the way I look and more leg strength. (133/M29/UK)

As we see, these responses are very similar to those given by female polers despite the differences in teaching, marketing and perception.

Ambivalence v. encouragement

Many polers would encourage, or at least countenance, men doing pole (for example, KT said 'everyone should do it') but there is also some ambivalence because pole has been, almost entirely, run by women, with an all-female, feminised environment where women feel able to grow in physical confidence. Of course, as well, pole has no definitive 'leaders'; it is run along egalitarian lines, it prides itself on how it can improve women's lives and it encourages community and cooperation, so perhaps there is also wariness that too many men might change things. 'One might argue somewhat cynically that Western middle-class boys already have a multitude of recreational choices and that there is hardly a pressing need to colonise traditionally female pursuits' (Gard, 2001, p. 223) which Megan echoed:

> In general we find that it's gay guys, you know, which would be OK but still, you know, it's a man in the class whether he's gay or not, so yeah, I don't do classes for men. Men can go and do all kinds of other shit [sports, etc.], you know.
>
> (Megan)

Similarly, Michael Atkinson (2008, p. 72) points out:

> Men's growing interest in cosmetic surgery might empirically hint toward the emergence of a late modern 'male femininity'; a gender status that at once draws on and seeks to reaffirm traditional images of men and the power bases men hold, but also tactically poaches and re-signifies stereotypically feminine symbols and practices in order for their male deployers to appear as progressive, neo-liberal and socially sensitive.

What may save pole from being 'poached and re-signified' is that the very rhetoric, such as friendship, fun and empowerment, which encourages women to pole (whether high or low) can simultaneously work against men doing it, as does its history.

> What some of them really want is a mixed class which I won't do because we talk about, oh you know, our bras and periods, I know when they're on their periods and stuff like that, so we won't be talking about that if there's a guy in the class. And they would get more competitive.
>
> (Genevieve)

Although pole does get more competitive at advanced levels, and between instructors of high profile schools, in general the classes eschew a competitive spirit in favour of a supportive approach, as we saw in Chapter 7. So the participants, as we also saw above, expressed ambivalence:

SH: When men come do they say what they are there for?

BOBBI: A lot of the guys, the ones that turn up in dresses and trying to find a place to hang out are dancers and they want to learn another kind of dance, they're actually professional dancers already and have heard about pole dancing strength and the help with their flexibility, etc etc, in which case I'm fine with that.

SH: So do you find that it's different with men, because you know with their body strength for example. Do they build strength?

BOBBI: They pick up much easier because they're so strong, they do things that girls spend weeks and weeks doing and still can't, but they haven't got the grace to be honest, the whole package isn't right, it doesn't look right, its just, to look at, no matter

> how good a guy is on a pole, it just doesn't fit right, if they want to do it great, but I wouldn't go to specifically watch a man on a pole. ... But good luck to them.

In general, there are still negative perceptions about men and boys as dancers and pole is still, in many imaginations, such a feminised and sexualised activity that it is questionable whether it will appeal. Gard (2001) discusses a children's story called 'Jump' in which a little boy is told that 'real boys don't go to dance classes'; eventually he becomes the best dancer in his class. One of the respondents said that he would like to teach pole but his family were very against his involvement so he could attend only occasional pole jams, when the opportunity presented itself, often covertly. He would like for pole to be accepted as exercise and for it to be included in the Olympics, because then his family might be more willing to support him doing classes: 'hope it can for the 2012 but still a long while to wait'. As Atkinson (2008, p. 69) argues,

> [f]ew have studied, for instance, how 'everyday' men engage body-work in order to appear 'regular', or have responded to broader cultural fluctuations in masculine hegemony with scripted body ritual. Fewer still have inspected how men play with innovative forms of aesthetic masculinity (i.e. beyond the context of 'gym work', tattooing or other stereotypically masculine body projects).

If pole for women becomes about feelings of empowerment or achievement, of improved confidence, of having fun and making friends, then for men it is at least some of those things; it also functions as a sort of resistance to hegemonic, body building identities, replacing them with 'aesthetic' masculinity which is still about strength. This is important because, as David Morgan (1993, p. 70) points out,

> [a] somewhat one-dimensional picture of men and their bodies emerges, one over-concerned with hardness, aggression, and heterosexual performance. ... In almost any society, it is relatively easy to think of those sites or arenas where men's power, expressed in bodily terms, is exercised. In most modern societies, for example, we might routinely think of the sports arena or the battle-field (ibid. p. 76)

Men's pole has become associated with hardness and strength; as we saw above, it appeals to rugby and parkour players, skydivers and climbers, all keen on perfecting power moves and building muscle. But they are not the whole story. Jonathan Watson (1998, p. 176), on his participants' accounts of men's healthy and unhealthy embodiment, notes that,

> such accounts contained critiques of and resistance to ideal images of mesomorphic or androgenous men. In doing so they appear to make space for the body-I-am, the body-I-have ('a normal everyday body'). Their own embodiment was most obviously located in the social spaces of marriage, fatherhood and work. In the contexts the emphasis was on function (having normal, pragmatic everyday bodies) rather than appearance.

In the same way we find that many men who enjoy pole classes, while currently relatively small in number, do find a way to an 'everyday' body which occupies neither of Atkinson's (2008, p. 68) 'polar positions' of either 'hegemonic/traditional' or 'drastically alternative and deeply marginalised', but 'neither of these polar positions accurately captures how clusters of men often wrestle with and negotiate established constructions of masculinity in novel ways' (Atkinson, 2008, p. 68). Instead it is slightly flamboyant; it is slightly showy (not too much to de-masculinise it); it is aesthetic but appropriately powerful. I would also posit another reason: I have often heard that men who started pole classes had a partner, friend, or relation who had themselves done a pole class or a hen party, and had heard how much fun it was. Although arguably adult men do have more outlets for a liberating physicality, obviously not all men do (as the oldest respondent demonstrated), so perhaps we should assume that men also enjoy the same feeling of physical liberation, as if they were swinging on a rope as a child, just as many female polers did. As one respondent said, the appeal was, for him:

> Spins. That simple.
>
> (133/M29/UK)

This was the final empirical chapter. In the next and final chapter I discuss the main themes which have arisen from the data and attempt to draw some conclusions about pole classes, embodiment, empowerment and pleasure.

Conclusion: A Positive Active Identity?

This book began with a description of a poler. Pole has multiple, contradictory meanings and its practice generates questions about display and objectification. But we have also seen that pole is a positive activity for many women all over the world; pole represents a complex intersection of body, pleasure, fitness and friendships. Pole classes, and their attendant associations and cultural anxieties which radiate outwards like ripples on water, leave me perplexed for a number of reasons. Mostly because I know that by this point I should, perhaps, posit several airtight conclusions and yet I am not going to even attempt to be so glib. As I warned in Chapter 1, this research has proved to be of the sort from which new questions proliferate rather than questions being answered. So I am not going to attempt to fool anyone into thinking that I believe there are definitive answers to a phenomenon like pole classes: I do not. It will always divide opinion. This is not to say we should lazily avoid discussion; only that sometimes we have to acknowledge that human actions often cannot be tied up in a neat bow. As I also said in Chapter 1, qualitative research is not an exact science – but it does offer us an exquisitely detailed and current window on our world.

This book has attempted to knit together various disparate threads: embodiment, ideas about empowerment, physical activity and physicality, femininity and feminism, body image and pleasure. I have only begun to scratch the surface of the complexities of pole classes in relation to agency and pleasure. What is going on here? We have women willing to pay more money than at an average exercise class, willing to attend week after week when previously most of them had disliked exercise, willing to keep trying a difficult activity which left them bruised and sore, willing to attend despite time and family responsibilities and moreover, willing to engage in an activity often associated with clubs in which women are objectified. Some of the main themes which arose from both the interviews and the questionnaire were of camaraderie, friendships and respect for instructors. The women had fun together, whether dressing up in 'high' classes or concentrating on the exercise aspect in a 'low' class. Overwhelmingly, the wider pole community has a sense of being by and for women, a DIY, egalitarian ethic. Individual subject positions may have differed between strippery or exercise polers but overall they all saw themselves as part of a broader pole community. Analysis of pole classes cannot overlook its links to the sex industry. It is not whether a pole is incongruous in an exercise class but whether a pole is incongruous in a lap-dancing club, because it actually looks like a piece of exercise equipment. Wherever it is, it is, of course, just a metal pole which we choose to load with cultural anxieties, and its inscription depends very much on the individual. So if we cannot ignore its ties to lap-dancing clubs then, by the same token, we cannot ignore the consistent narratives of transformation, jubilation, increased confidence and exuberance. My sympathies have accrued over time; my position is grounded in my own experiences of the classes and in the positive accounts of the women I met and in my own feminism which (as I have said elsewhere, 2004, p. 3) has 'provided both method and motivation for my work'. I have come to agree with Martha McCaughey (1997, p. xii) who recommends (about self-defence but arguably also relevant here) that 'feminism take seriously the corporeality and pleasure of that resistance. It demands that feminism get physical'. As I have worked on this project I have also increasingly leaned towards what Catherine M. Roach (2007, p. 25) calls a 'sex positive' feminism, by which she means a 'type of feminist response ... [which] often also defend[s] pornography and prostitution as at least *potentially* empowering and legitimate, under the right circumstances, for the women who *choose* these professions [my emphases]'.[1] Or, as Lesa Lockford (2004, p. 89) explains, 'a sex-positive feminism challenges several assumptions that operate within more academically valorized feminist viewpoints'. This is not always the most comfortable position to operate from, so this chapter will perhaps also chart how I reached it.

Decency and agency

Many of the participants and questionnaire respondents expressed disappointment, frustration, shame or anger over the attitudes of others towards them for attending pole classes. These 'others' ranged from feminist academics who assumed that classes were somehow an offshoot of the sex industry; to journalists who were interested only in a salacious angle to titillate their readers before they moved onto the next story; and to friends or relations who saw pole as being somehow demeaning or shameful. One example is Ariel Levy (2006, p. 34) who writes:

> We have to wonder what women are getting out of this now. Why would a straight woman want to see another woman in fewer clothes spin around a pole? Why would she want to be on that pole herself?
>
> (Levy, 2006, p. 34)

These questions lack logic and are typical of objections to pole classes by people who have never actually attended one or spoken to polers. If we were to extrapolate Levy's logic, why watch a ballet dancer or a gymnast, or indeed attend a gym at all? Overall, many polers felt embattled by this kind of misconception and wished to change the image of pole. Commonly, women's agency is seen to be minimal or illusionary because women exist within power structures which can only contain them. But women do find ways to achieve agency within those structures (Wearing, 1995) and I return to this a little later. In feminist debate, women's sexual agency has commonly held a see-saw position: sexual liberation advocated and feared, its repercussions positive for women while also holding them in the bind of patriarchy. As Amy Wilkins (2008, pp. 330–31) argues,

> [w]hile some feminists applaud women's increasing sexual agency, others argue that changes in sexual expectations have only increased the pressure for young women to engage in sexual behaviors they might not otherwise choose (Jacobs Brumberg 1997; Pipher 1994). A discourse of victimization thus pervades discussions of adolescent girls' sexuality. This discourse, which positions young women as passive recipients of unwanted sexual attention or as pressured into early or more frequent sexual behavior acknowledges girls' relative disempowerment in heterosexual interactions but precludes any discussion of sexual desire on the part of young women. In an article aptly subtitled 'The Missing Discourse of Desire', Michelle Fine (1988) noted that girls' voices of desire, submerged under this discourse of victimization, are glimpsed only fleetingly by positioning girls as victimized rather than desiring subjects.
>
> (Wilkins, 2008, pp. 329–30)

Pole remains sexualised in the public imagination. This fact is probably irrefutable and probably also fixed. However, pole goes beyond being merely a sexualised activity where women are 'passive' victims of attention, it accepts (in fact, often creates) women as agents of sexual desire or desirability (whatever their age or size), and within the pole community women are active subjects rather than passive objects, because the pole community is run mainly by women who champion pole as an activity suitable for all women.

Embodiment

As Young (1990, p. 66) argues, women internalise gender norms about their embodiment:

> The woman lives her body as object as well as subject. The source of this is that patriarchal society defines woman as object, as a mere body, and that in sexist society women are in fact frequently regarded by others as objects and mere bodies. An essential part of being a woman is that of living the ever present possibility that one will be gazed upon as a mere body, as shape and flesh that presents itself as the potential object of another subject's intentions and manifestations, rather than as living manifestation of action and intention.
>
> (Young, 1989, p. 66)

Arguably, pole classes – admittedly limitedly, within the classes – offer possibilities for 'action and intention' as well as offering an opportunity to be more than an objectified body which is gazed upon. Many of the women I spoke to overwhelmingly had a history of not liking exercise and of having previously 'failed' at it. They were in the habit of defining themselves apologetically or negatively as clumsy, or not physical, not fit, not strong, or not sporty because they did not, and probably had never, enjoyed the sorts of physical activities that are traditionally seen to be acceptable. Nonetheless these same women said they had seen benefits in their fitness, strength and body confidence, both during and after the classes. They had found, in the classes, opportunities for 'action and intention'. As Jan Wright and Alison Dewar (1997, p. 80) point out, discourses about women and exercise often 'had few resonances with our own experiences of physical activity nor did they seem to take into account the movement of the body as a source of the kinaesthetic/sensual pleasures which we described to each other'. The participants of this study expressed surprise and enthusiasm about enjoying some physical activity (although they were guarded about whether it would translate into a new-found love of physical activity per se).

Pole does challenge discourses about women's bodily experiences: it challenges the 'restricted movement [and] stationary positions' (Adair, 1992, p. 41) that femininity demands. It resists the discourses which tell us that a woman tends 'not to reach, extend, lean, stretch, and follow through in the direction of her intention' (Young, 1990, p. 146). When doing pole a woman cannot help but learn how to reach, extend, lean, stretch and follow through. She also learns, among other physical skills, to climb, to swing, to hold her own body weight, to balance and to invert. She encourages other women to grow in strength and confidence. A pole body may be lightly muscled but it is strong. It is not a static body either, it is creative and confident – all the things that we deplore as lacking, for women's bodies, in cultural discourses and narratives. As Martha McCaughey (1997, p. 166) has noted,

> [f]eminists have focused less on how the body might be a source of new consciousness – in short, on the body as something other than a passive house of the soul. The body is not just something that gets acted on, taken over, occupied, constrained or defended. Feminists might conceive the body as an agent, not just the thing that (physical) agents struggle over.
>
> (McCaughey, 1997, p. 166)

Yet we have seen that pole offers women a way to think of their bodies as something other than passive or constrained. Similarly, as Tara Brabazon (2006, p. 69) points out, 'the medicalisation of femininity is a result of patriarchal hierarchies in sport and the workplace which label women inferior, sick, weak or vulnerable' and 'feminists have a clear task here: we must critique and counter – at every opportunity – this pathologisation of women's flesh' (ibid., p. 74). If pole is not an appropriately feminist activity, even while it is primarily practiced, taught, run by and watched by women, then what is? Finding agency in this way is, by some of the participants, read as feeling empowered. Empowerment, discipline and pleasure in the body are all prevalent discourses around women re/discovering physical activity (for example, see Gilroy, 1997; Wright and Dewar, 1997). As we saw several participants used the term 'power' or 'empowerment' but not, as Wright and Dewar (1997, p. 91) explain, in the hegemonically masculine way that 'power' is understood and which implies aggression or violence. Rather, their empowerment was about the 'setting and achieving of personal challenges ... being in control; identification with body and pride in its/their achievements' (ibid.). Power was, in this case, personal and embodied. Many of the women who had previously dropped out of exercise classes began to attend because the class became a place to socialise as well as to 'exercise' rather than only because they wanted to try to get fit. As Segar et al. (2006) found, women who exercise for non-body-shape reasons are more likely to continue attendance. Segar et al.'s study didn't report women who change from one category to the other but, in the pole classes, it seemed that even those who started out in the body-shape category could move over to the non-body-shape category thus ensuring that the chances of them not dropping out were much improved.

Empowerment?

Andrew Sparkes (1995, p. 116) recommends that we examine the 'central moments or critical incidents' in our data, that we look at the 'indecision' and the 'contradiction' in order to arrive at a 'more complex view of reality'. The polers negotiate a number of complex processes in order to both distance themselves from feminism and to utilise its gains. Individualism, with its mantra of 'I achieved this', does have its benefits in that the polers utilised it to negotiate access to, and motivate themselves to attend, classes (as I discussed in Holland, 2009). But, as Rich (2005, p. 501) points out,

> the narratives ... tended to assume a separation of self from gendered contexts, and a belief that ones individual determination to become or achieve something ... is enough to overcome social constraint. ... Indvidualism rather problematically obscures underlying social structures. It assumes that as a detached self one is in a position of independence.

So we see that there is a tension between individualism (where there is no collective gain but there is personal gain which may then be used to help others), the distancing from feminism and the utilisation of feminist gains and objectives. Added to these tensions are those created by the difference of approach between 'high' and 'low' classes. Polers occupy more than one subject position, drawing on various discourses such as autonomy, physical strength, independence, sexiness, fitness and their overall identity as a poler. They are often guided by feminist ideals. Pole comes to be seen by many polers as more than an exercise class; it comes to represent an unmooring from constraint, from the 'critical incident' be that weight gain, boredom or isolation, or from a long-standing aversion to hegemonic exercise forms which held for them negative connotations. For instructors it was slightly different: the discourse of individualism, of 'I did this', was tempered by their zeal to convert non-polers and assist existing polers to remain so. However, a further complexity is that a discourse of individualism, coupled with the unmooring mentioned above, also resulted in an inevitable binary, that of 'high' or 'low', which divided just as much as it drew together.

A major contradiction within pole classes, which can be a source of frustration for those with 'exercise' subject positions, and which is the root of the mixed messages discussed in Chapters 5, 8 and 10, is that 'strippery' classes appear to tie polers to its origin in lap-dancing clubs. But the experience of the exotic dancer is not always as straightforwardly negative as we may at first assume, and as Catherine M. Roach (2007, p. 13) pointed out:

> The answer to whether stripper is 'demeaning' or 'empowering' will obviously differ for each dancer, often on a night by night basis, depending on her reasons for entering the industry and her experiences in it. Even if we understand these two terms as opposite ends on a continuum of evaluation, such an inquiry would still miss the complex, ambiguous and often contradictory realities of women's work and lives as dancers. 'Demeaning or empowering' is, like any dualistic question about a multifaceted area of human experience, impossible to answer and in fact misleading as posed, although it's a common enough question on the talk-show circuit. ... I don't believe that stripped can be labelled in a black-and-white fashion as either necessarily degrading or as potentially liberating.

Rambo et al. (2006, p. 221), writing about strippers, argue that 'scholars have regulated, disciplined and controlled female bodies and selves through the discourse of commodification' and that through this regulation they have refused to listen to any alternative narratives where a woman may profess enjoyment or empowerment, instead she is 'characterized as a victim of false consciousness – a passive agent and cultural dupe who has internalised her own oppression' (ibid., p. 217). Exercise polers wish to disassociate themselves from the context of lap-dancing clubs – which strippery classes constantly re-tether them to – because 'resistance against female sexualized display is vehemently grounded in the desire of women to be accepted as thinking individuals and not, as Griselda Pollock stated in1981, "explicitly as cunt"' (Willson, 2008, p. 7), and because of the regulation highlighted by Rambo et al. However, sexuality need not (indeed, must not) always be seen as negative, otherwise we also would be guilty of defining women only as passive, sexualised beings. As Amy Wilkins (2008, p. 346) found:

> Goth women experience their sexuality as personally empowering: It provides them with a sense of control over their bodies, with the right to feel and act on desire and with external validation of their expressions of sexiness. For women struggling to walk the narrow sexual line mandated by the mainstream culture, these gains should not be understated. They are mitigated, however, by the persistence of sociocultural ideas that position men as sexual consumers/owners. As feminists have argued about the sexual revolution, simply increasing women's right to enjoy sex does not undo the basic heterosexual relationship that confers men with sociocultural power. Indeed, in the absence of other changes, women's sexual freedom benefits men more than it does women by providing men with greater sexual access to women without altering heterosexual power arrangements.

Wilkins's goth participants, just as Roach and Rambo et al. found with the exotic dancers, wanted to feel sexy and sexual, but were limited by societal power imbalances. However, they were able to feel personally empowered and were in control of their bodies; this individualisation echoing that reported by the polers. Within power constraints, then, women can find room for personal empowerment, if not empowerment for all.

Appropriate bodies

Pole classes deliver gendered embodiment. By this I mean that, while developing the poler's physical strength and confidence, it also doesn't 'butch up' the poler; her muscles are toned and lengthened. Strippery classes add to the sense of appropriate embodiment in that the polers wear stripper shoes and their body rolls and other linking moves are consciously sexual (although some would say sensual, drawing a fine distinction between the two). Pole is not contentious and fits into models of women being non-competitive; this is true in the classes. Advanced polers are competitive and the fierce competition and high level of skills displayed during pole events testify to that, but most of them also have a history of playing sports which instil a sense of competitiveness. However, it also gives the poler, at any level, a sense of physical liberation and confidence which was alien to them as adults (particularly the non-exercisers); they referred repeatedly to the feeling of liberation they felt as children when playing on a rope or monkey bars. Jan Graydon (1997, p. 68) argues that self-confidence is 'linked with continuing sports involvement ... as well as with persistence in physically demanding tasks'. This rediscovery of the fun and physical liberation of childhood has led to many pole schools offering other forms of dance and exercise: for example, can-can, hula-hoop and cheerleading classes. Some schools offer related classes to capitalise on pole's success and which would improve the poler's body strength and flexibility: for example, Bobbi's runs pole gymnastics, and Studio Verve runs athletics courses; schools in the UK run Chinese acro-pole classes.[2] In these ways, pole can be seen as a form of resistance to male-dominated activities.

Despite this it is not, obviously, without its contradictions and 'high church' classes offer the most obvious tensions. Susan Bordo (1989, p. 22), writing about eating disorders, argues that 'the pathologies of female protest function, paradoxically, as if in collusion with the cultural conditions that produce them, reproducing rather than transforming precisely that which is being protested'. Stripper shoes, for example, remain one of the main links to lap-dancing clubs. They reproduce a certain look and many women enjoy wearing them, even though they are not comfortable; they offer a way to feel desirable, they are a way to play with an image. Yet the shoes seem almost a way to hobble oneself, to handicap in sporting terms as well in general physical terms, while at the same time conferring height and confidence. Many of the participants saw even being able to walk in the shoes as an achievement in itself and somehow great fun, while many schools warned 'this is exercise so no heels'. A recent article (Copping, 2009, p. 44) in the *Financial Times Weekend Magazine* revealed that there are now £90 (around U$150 and AU$179 at time of writing) three-hour workshops in London, at which a 'vertiginous footwear guru ... in towering Perspex strippers shoes ... instructs stiletto-wearing virgins and amateurs alike how to wear high heels'. Using techniques in improving posture and coordination the author notes that the class is 'as much about "go-girl" confidence-boosting as it is about not falling flat on your face'. The end of the lessons sees the student 'parading along a busy West End road, showing off her new-found talents. Excruciating it certainly was but, I admit, also rather liberating' (ibid.). The liberation came from the stares she got from taxi drivers and hotel bellboys. Yet walking a street in shoes that are a cross between wearing stilts and being hobbled, of

course, is not liberating; such shoes do not only hamper running or walking quickly and easily, they actually prevent it. Shoes which are so closely and universally associated with the sexualisation and availability of women's bodies cannot, by any stretch of logic or imagination, be liberating. Certainly I accept the author may have had fun at her lesson and be pleased with her new balancing 'skill'. I felt dismayed by the article because that kind of contradiction does not benefit the perceptions of pole classes and is why we must tease out the differences between 'high heels workshops' and pole classes; why we must not simply and lazily lump everything in together and label it 'sexualisation of culture' or 'false consciousness'. Stripper shoes are not, and were never, meant for everyday wear, but because of celebrity culture we see more and more photographs of 'celebrities' in such shoes, and because of the pornification of some aspects of culture, we also see them posing in teeteringly high shoes (posing, note, not walking further than to a waiting car) and this has resulted in them being seen as the norm. Groups of young women totter around in them on nights out.[3] But even 'high church' instructors did not advocate them as everyday wear: they are a prop, part of a performance; they alter a poler's posture while also making her feel taller, even Amazonian; perhaps they even complete her stage persona and make certain moves more possible or more dramatic. What makes them appeal for performance is what also makes them impractical for anything else; for example, Suzie Q said they hurt and Bobbi said she still wondered if she would hurt her ankle in them. No one is expected to leave a class in them and walk to their car or bus stop.

Finally

For me, pole classes are a wonderful thing for those women who find the classes transformative, and a beautiful awe-inspiring thing to watch. I still have reservations about lap-dancing clubs and concerns about sex workers who may or may not be choosing their occupation from a limited range available to them. But (like Lesa Lockford, Jacki Willson and Catherine Roach) these reservations were tempered by my meetings with Bobbi, Suzie Q and others; and by my reading of, for example, Carol Rambo and Elisabeth Eaves. This chapter, then, has explored the key themes which arose; the book has contributed to debates about empowerment, physicality, femininity and pleasure; and hopefully it has created further questions. Pole classes may have originated in lap-dancing clubs but they do not take place in lap-dancing clubs and their context differs. Several polers, especially instructors (such as Genevieve,

Megan and Bobbi), wish to call pole 'pole dancing' because it best indicates that it is an art form, a skilled physical activity. Others (such as Evie and Jennifer) would prefer that it not be called pole dancing only because they wish for classes, and performance, to be recognised as an evolved and evolving activity separate and distinct from the context of lap-dancing clubs. Pole is performed primarily by women, just as it is in lap-dancing clubs. Several participants refuted that pole dancing in a lap-dancing club was disempowering for them: Bobbi loved the work and missed it; Suzie Q had bought property with her earnings. Others were ambivalent: most notably Silke, who was wary of telling her boyfriend because she believed he would disapprove. Valerie Walkerdine (2003, p. 242) argues that there is a 'long-established incitement to women to become producers of themselves as objects of the gaze'. But pole classes are not performed for male spectators; nor, therefore, are the performances predicated on sexualised display. Indeed, one of pole's most positive aspects is its feminised, female environment which fosters camaraderie and encouragement. Historically, many dance forms have suffered initially from stigma, such as ballet, the waltz and salsa, but over time have become accepted as an art form. Some may argue that this process of normalising actually represents the insidious way that the previously unacceptable becomes acceptable, but we must also bear in mind that art forms, of all kinds, will always evolve, respond to the current culture and society and will challenge current ways of how we view creative endeavours. Pole may look the same in a lap-dancing club as it does in a class but, again, the reasons for it being performed differ: women attend pole classes for fun, for friendship, to get fit, for feelings of achievement and pleasure. Certainly it is the same, but it would be a mistake to assume that pole classes occur in the same context. Rambo et al. (2006, p. 224) argue that women should be allowed the space to develop a positive, active physical identity, where they enjoy both their bodies and the attention. And, I would add, where we respect their choices enough not to judge or belittle them. As Judith Okely (1996, p. 214) argues, 'through careful examination and in the telling, we can discover that specific moments in individual lives inform us about both dominance and points of resistance'. Looking at pole classes prompts us to unravel a skein of comfort, pleasure, hard work and fun, and we must remember that pole classes differ in context, not only from lap-dancing clubs but often from each other, and they all offer benefits to those who do it. We also then see its often hidden points of resistance.

Appendix 1

Questions for Online Questionnaire and Statistics of Responses

1. Are you	Response per cent	Response count
Female	**97.8%**	132
Male	2.2%	3
	Answered	**135**
	Skipped	**0**
2. What is your age? (exact or ballpark)		**Response count**
		135
	Answered	**135**
	Skipped	**0**
3. How would you describe your ethnic group?		**Response count**
		117
	Answered	**117**
	Skipped	**18**
4. What is your status?	**Response per cent**	**Response count**
Married/civil partnership	22.2%	30
Co-habiting	26.7%	36
Single	**30.4%**	41
In a relationship but not living together	19.3%	26
Other	1.5%	2
	Answered	**135**
	Skipped	**0**
5. How many children do you have?	**Response per cent**	**Response count**
0	**70.4%**	95
1	14.8%	20
2	10.4%	14
3+	4.4%	6
	Answered	**135**
	Skipped	**0**
6. Where do you live? (Please name just your town & state/county & country)		**Response count**
		135
	Answered	**135**
	Skipped	**0**

(continued)

Continued

7. What is your occupation?		**Response count**
		135
	Answered	**135**
	Skipped	**0**
8. Have you always exercised regularly?	**Response per cent**	**Response count**
Yes, always loved exercise	**40.8%**	53
No, usually hate exercise	24.6%	52
On & off	35.4%	46
	Other (please specify)	13
	Answered	**130**
	Skipped	**5**
9. Would you say your body image was	**Response per cent**	**Response count**
excellent (I look good)	21.5%	29
OK (it could be worse)	**38.5%**	52
Up & down (I have good and bad days)	37.8%	51
Generally bad (I hate my body)	5.2%	7
Not sure	0.0%	0
	Answered	**135**
	Skipped	**0**
10. Please say what OTHER exercise classes you have attended in the past or currently attend?		**Response count**
		125
	Answered	**125**
	Skipped	**10**
11. Please say how long you have been doing pole classes?		**Response count**
		134
	Answered	**134**
	Skipped	**1**
12. How did you hear about them?		**Response count**
		133
	Answered	**133**
	Skipped	**2**
13. Are you	**Response per cent**	**Response count**
a pole instructor	23.5%	31
a pole student	**64.4%**	85
neither	4.5%	6
both	12.1%	16
	Other (please specify)	18

(*continued*)

Continued

	Answered	**132**
	Skipped	**3**
14. Where do you do the classes? (e.g. pub, pole studio, etc.)		**Response count**
		134
	Answered	**134**
	Skipped	**1**
15. Why did you start to do the classes? What attracted you to pole?		**Response count**
		134
	Answered	**134**
	Skipped	**1**
16. Do you, or have you previously, worked as a professional pole dancer in clubs?	**Response per cent**	**Response count**
Yes I did	9.2%	12
Yes I do	1.5%	2
No	**85.4%**	111
I plan/hope to be	4.6%	6
Prefer not to say	0.0%	0
	Other (please specify)	17
	Answered	**130**
	Skipped	**5**
17. Do you ever have difficulty paying for your lessons?	**Response per cent**	**Response count**
Yes	14.4%	17
No	**67.8%**	80
Sometimes	17.8%	21
	Other (please specify)	21
	Answered	**118**
	Skipped	**17**
18. How many women/men attend your class?		**Response count**
		132
	Answered	**132**
	Skipped	**3**
19. What do you particularly enjoy about pole classes?		**Response count**
		128
	Answered	**128**
	Skipped	**7**

(continued)

Continued

20. Do you intend to continue to attend pole classes indefinitely?	**Response per cent**	**Response count**
Yes	**84.5%**	109
No	3.1%	4
Not sure	12.4%	16
	Answered	**129**
	Skipped	**6**
21. What are the positive things that pole classes have given you (e.g. improved body image, self confidence, body strength, overcome shyness, etc.)?		**Response count**
		132
	Answered	**132**
	Skipped	**3**
22. Any there any negative things?		**Response count**
		120
	Answered	**120**
	Skipped	**15**
23. Do you have, or would like, a pole at home?	**Response per cent**	**Response count**
Yes	**74.2%**	98
No	2.3%	3
Soon	7.6%	10
One day	16.7%	22
Not sure	1.5%	2
	Other (please specify)	23
	Answered	**132**
	Skipped	**3**
24. Would you, or have you, used DVDs at home to teach yourself?	**Response per cent**	**Response count**
Yes	56.3%	72
No	24.2%	31
Maybe	18.0%	23
If I could afford them	1.6%	2
Not sure	0.8%	1
	Other (please specify)	22
	Answered	**128**
	Skipped	**7**
25. Why do you think pole classes are so popular?	**Response per cent**	**Response count**
Something different	87.1%	115
Fun	**88.6%**	117

(*continued*)

Continued

Sexy	61.4%	81
Making friends in class	56.8%	75
Good instructors	48.5%	64
For fitness/weight loss	73.5%	97
	Other (please specify)	15
	Answered	132
	Skipped	3
26. Is there anything else you would like to add about your experience of pole classes?		**Response count**
		67
	Answered	**67**
	Skipped	**68**
27. If you would like me to interview you (UK 2008; or Sydney March 2008; or NY June 2008) please put your e-mail address in the box below and I will contact you. Thanks again. Best wishes, Sam		**Response count**
		65
	Answered	**65**
	Skipped	**70**

Appendix 2

Websites of Instructors Who Took Part + Schools Mentioned

Websites of instructors or schools who asked me to include a link:

Aussie Pole Men http://www.mangodance.com.au/Mango%20Dance%20Productions/mango_troupe_AUSSIE%20POLE%20BOYS.html
Bobbi's http://www.bobbispolestudio.com.au/
Holistica http://www.holisticahealth.co.uk/
Miss Suzie Q http://www.ipole.com.au/home.html
Pole-da-Cise http://www.pole-da-cise.com/
Pole Dancing School http://www.poledancingschool.com/
Polepeople http://www.polepeople.co.uk/
Studio Verve http://www.studioverve.com.au/
The Flying Studio http://www.strictlypoledancing.co.uk/
Vertical Dance http://www.verticaldance.com/
Zebra Queen http://www.zebraqueen.co.uk/
Jason Parlour photography http://www.jasonparlour.com

Websites of other schools or instructors mentioned in the text:

Art of Dance http://www.theartofdance.co.uk/
ETEDance http://www.empowermentthroughexoticdance.com/
Fawnia Mondey http://www.officiallyfawnia.com/
Felix Cane http://www.felixpoledancing.com.au/
International Pole Federation http://www.polefederation.com/
Jamilla DeVillle http://www.jamilla.com.au/
Kiwi Pole http://www.kiwipolefitness.co.nz/
Pantera Blacksmith http://www.panteraspoleworld.com/
Pole Dance Community http://www.poledancecommunity.co.uk/
Pole Diamonds http://www.polediamonds.co.uk/
Pole Dolls http://www.poledolls.co.uk/
Pole Love http://www.polelove.co.uk/
Pole Passion http://www.polepassion.com/
Pole Stars http://www.polestars.net/
Purity http://www.puritypoledancing.com/
S-Factor http://www.sfactor.com/
Spin City Pole Fitness http://www.spincitypolefitness.com/
UK Amateur Pole Performer http://www.ukamateurpoleperformer.com/
World Pole Dance http://www.worldpoledance.com/
YUPE http://www.yupe.co.uk/

Notes

Introduction

1. There is even a pole studio called Shanyn's Artistic Strength Studio in Ontario, Canada.
2. Available on YouTube.
3. 'Alternative' women are those who are, or have been, part of a subculture and whose appearance is non-mainstream, for example, multiple tattoos or piercings.

1. Towards a Feminist Ethnography

1. Unfortunately, the recording of Elena's interview was unclear in many places because of the noises of the café where we met.

2. From Circus and Sex ...

1. This situation is now changing: 'In October 2009, Helsinki will see something it has never seen before: an international pole dance competition with performers and judges from Sweden, Denmark, Finland, Russia, the UK, France, and Australia. ... This is somewhat surprising for a country that didn't even know pole dance existed until November 2007 when the first pole dance school, Rock the Pole, opened its doors in Helsinki. Soon after this two other schools, Finnish Pole Dance Ltd. and Pole-a-holic emerged in Tampere and Seinäjoki. Now, there are a handful of larger and smaller schools around the country' (from http://www.poledancecommunity.co.uk/articles/articles/pole-art-2009-pole-dance-takes-over-the-north.html, accessed on 14 July 2009).

3. ... to Fitness and Leisure

1. Also, some accounts say that Arabic dance was never a dance performed to mixed audiences; that women would perform only to other women to celebrate birth.
2. Indeed, the pole school Pole Diamonds in Yorkshire, UK has 'Alternative Fitness' as its strapline.
3. James's reply to my question 'do you think that being a professional dancer does help you?' concurs with Lilia's assertion:

JAMES: Absolutely, with the flexibility.
SH: Artistically?
JAMES: Artistically yes because being a dancer you know how to perform with your body already. And that is half the battle to

when you are actually doing a show, otherwise it does look like someone's just doing their tricks up against the pole, there is no performance there at all.

4. What is a Pole Class?

1. In fact, a respondent to the questionnaire said that she first heard about pole classes when she 'saw an advert in the *Metro* newspaper asking "have you ever seen a fat pole dancer?"' (58/F35/UK).
2. On the Studio Verve website PoleFit or PoleFitness is always followed by a ® symbol denoting that it is a registered trademark.
3. Other forms include, as mentioned in Chapter 2, pole used within a burlesque performance; or pole used in a circus performance.
4. I return to Pole-Da-Cise in Chapter 9.

6. Diversity and Empowerment?

1. Another example is that there are two Empowerment Through Exotic Dance, Ltd. (ETEDance) studios in Illinois, US, owned by Mary Ellyn Weissman, who I mention again in Chapter 8.
2. For a longer account of the incident go to http://artofdancesam.blogspot.com/.

7. 'A Thing of Beauty'

1. In fact, there is a group on Facebook called 'I Pole Dance & Here's The Bruises To Prove It!'
2. The Pole Dance Community website now sells ankle protectors to improve grip but particularly to protect the top of the foot.
3. A hen party or hen do is when a bride-to-be and her group of friends and family (usually all women) have some kind of social event to celebrate the bride-to-be's impending wedding. It is also known as a bachelorette party, or even stagette.

8. The Pole Community: Opening Closed Minds

1. Mary Ellyn is the owner of Empowerment Through Exotic Dance and perhaps one of the oldest celebrity polers at 49.
2. It is Sheila Kelley of the S- [or Strip] Factor who many people cite as being how they found out about pole because she was on the Oprah show.
3. These are just examples; there are many more 'famous' polers, some who do workshop tours, some who don't. There are also famous polers who haven't quite reached 'superstar' status just yet.
4. It sold 88,000 copies in the UK, double the sales of the Pirelli calendar it had set out to copy.
5. Organised by Pole Passion in Brighton, UK.
6. However, personally, I do not believe that removing a pair of ballet slippers constitutes stripping. The video of Elena's performance is available on YouTube.

7. Maypole dancing is a form of folk dance from Western Europe. Traditionally May Day, particularly until the early seventeenth century, was when people would dance around a beribboned, decorated pole; people would decorate their houses with flowers; and a May Queen would be crowned.

9. Case Study 1: 'Empowering Women with Confidence'

1. Unfortunately, the credit crunch has taken its toll, as Libby told me in an email in June 2009: 'we run classes 4 days a week now, Fridays have gone and Saturdays, and rather than doing 7 classes on a Sunday I now only [teach] 3'.
2. At the time of my visit they also intended to begin running workshops and classes for women who had suffered from cancer, offering advice on healthy eating and lifestyle from experts and tasters of pole classes for those who felt able to try some exercise.
3. BBC television series for budding entrepreneurs.

10. Case Study 2: Power Moves and Everyday Bodies

1. Set up by Lucy Misch who went on to set up the school Pole Exercise (which has the strapline 'pole dancing evolved'). From her website: 'Lucy set up the UK's first Pole Exercise Club. She had a vision – now shared by many – of bringing pole dancing out of the nightclub and into the living room, knowing what great exercise (and great fun!) it could be. She was also one of the first to target this unique form of exercise at men as well as women, over time developing a distinct style of pole dancing with them' (http://www.lucymisch.com/biography/, accessed on 20 July 2009).
2. The owner of Kiwi Pole does concede that pole hasn't yet properly 'caught on' with men in New Zealand because its image is still too feminised.
3. Parkour is a sport, originating in France, in which practitioners move from one place to another, usually outdoors and without spectators, getting over obstacles (such as buildings!) as efficiently as possible.

Conclusion: A Positive Active Identity?

1. I should also point out that this shift began several years ago during a roundtable discussion I had organised at my place of work. A representative from the English Collective of Prostitutes challenged a 'sex-negative' academic by saying 'remember, it *is* only sex and if the woman chooses to do it what right do *you* have to judge her?'
2. Reebok has developed Jukari ('fit to fly') classes which are a cross between an aerial hoop and pole.
3. In fact, in 2007, Leeds City Council in the UK offered New Year's Eve partygoers in the city centre flip-flops and plasters to lessen the numbers of women who might have ended up in the Accident and Emergency section of the hospital due to injuries because of their shoes.

Bibliography

Aalten, A. (1997) 'Performing the Body, Creating Culture' in *Embodied Practices: Feminist Perspectives on the Body*, K. Davis (ed.) (London: Sage).

Aalten, A. (2004) 'The Moment When it All Comes Together': Embodied Experiences in Ballet', *European Journal of Women's Studies*, 11 (2004), pp. 263–76.

Adair, C. (1992) *Women and Dance: Sylphs and Sirens* (Basingstoke: Macmillan).

Afshar, H., Franks, M., Maynard, M. & Wray, S. (2002) 'Issues of Ethnicity in Researching Older Women', *Growing Older Programme: Extending Quality Life*, newsletter 4, pp. 8–9.

Andall, J. (2003) *Gender and Ethnicity in Contemporary Europe* (Oxford: Berg).

Anthias, F. (2002) 'Beyond Feminism and Multiculturalism: Locating Difference and the Politics of Location', *Women's Studies International Forum*, 25 (3), May–June 2002, pp. 275–86.

Armstrong, G. (2001) *Fear and Loathing in World Football* (Oxford: Berg).

Armstrong, J. (2008) 'Researching DIY Grrrl (E)Zine Culture: A Methodological and Ethical Account' in *Remote Relationships in a Small World*, S. Holland (ed.) (New York: Peter Lang).

Arthurs, J. (2006) 'Sex Workers Incorporated' in *Feminism in Popular Culture*, J. Hollows and R. Moseley (eds) (Oxford: Berg).

Aspinall, P. J. (2002) 'Collective Terminology to Describe the Ethnic Minority Population: The Persistence of Confusion and Ambiguity in Usage, *Sociology*, 36 (4), pp. 803–16.

Atkinson, M. (2008) 'Exploring Male Femininity in the "Crisis": Men and Cosmetic Surgery', *Body & Society*, 14 (1), pp. 67–87.

Atwood, M. (2003) *Negotiating with the Dead: A Writer on Writing* (London: Virago).

Attwood, F. (2002) 'A Very British Carnival', *European Journal of Cultural Studies*, 5 (1), pp. 91–105.

Attwood, F. (2006) 'Sexed Up: Theorizing the Sexualization of Culture', *Sexualities*, 29 (1), pp. 77–94.

Attwood, F. (ed.) (2009) *Mainstreaming Sex: The Sexualisation of Western Culture* (Oxford and New York: I. B. Tauris).

Beagan, B., Chapman, G. E., D'Sylva, A. & Bassett, B. R. (2008) '"It's Just Easier for Me to Do It": Rationalizing the Family Division of Foodwork', *Sociology*, 42 (4), pp. 653–71.

Biscomb, K., Matheson, H., Beckerman, N. D., Tungatt, M. & Jarrett, H. (2000) 'Staying Active While Still Being You: Addressing the Loss of Interest in Sport amongst Adolescent Girls', *Women in Sport and Physical Activity Journal*, 9 (2), pp. 79–97.

Borden, I. (2001) *Skateboarding, Space and the City: Architecture and the Body* (Oxford: Berg).

Bordo, S. R. (1989) 'The Body and the Reproduction of Femininity: A Feminist Appropriation of Foucault' in *Gender/Body/Knowledge: Feminist Reconstructions of Being and Knowing*, S. Bordo and A. Jaggar (eds) (New Brunswick, NJ: Rutgers University Press).

Bordo, S. R. (1993) *Unbearable Weight: Feminism, Western Culture and the Body* (London: University of California Press).

Bott, E. (2006) 'Pole Position: Migrant British Women Producing "selves" through Lap Dancing Work', *Feminist Review*, 83, pp. 23–41.

Brabazon, T. (2006) 'Fitness is a Feminist Issue', *Australian Feminist Studies*, 21 (49), pp. 65–83.

Brah, A. (1996) *Cartographies of Diaspora: Contesting Identities* (London: Sage).

Brande, D. (1996) *Becoming a Writer* (London: Macmillan).

Bruns, A. & Jacobs, J. (2006) *Uses of Blogs* (New York: Peter Lang).

Budgeon, S. (2001) 'Emergent Feminist(?) Identities: Young Women and the Practice of Micropolitics', *European Journal of Women's Studies*, 8 (1), pp. 7–28.

Bury, R. (2005) *Cyberspaces of Their Own: Female Fandoms Online* (New York: Peter Lang).

Bury, R. (2008) 'Remotely-embodied Friendships in Female Fan Communities' in *Remote Relationships in a Small World*, S. Holland (ed.) (New York: Peter Lang).

Caudwell, J. (2006) 'Queering the Field? The Complexities of Sexuality within a Lesbian-identified Football Team in England', *Gender, Place and Culture: A Journal of Feminist Geography*, 14 (2), pp. 183–96.

Charmaz, K. (2005) 'What's Good Writing in Feminist Research? What Can Feminist Researchers Learn About Good Writing?' in *Handbook of Feminist Research. Theory and Practice*, S. N. Hesse-Biber (ed.) (London: Sage).

Cherny, L. & Weise, E. R. (1996) *Wired Women: Gender and New Realities in Cyberspace* (Seattle: Seal Press).

Cockburn, C. & Clarke, G. (2002) '"Everybody's looking at you!": Girls Negotiating the "femininity deficit" they Incur in Physical Education', *Women's Studies International Forum*, 25 (6), pp. 651–65.

Connell, R. W. (2005) *Masculinities*, 2nd edition (Cambridge: Polity Press).

Consaluo, M. & Paasonen, S. (2002) *Women & Everyday Uses of the Internet: Agency & Identity* (New York: Peter Lang).

Cooky, C. & McDonald, M. G. (2005) '"If You Let Me Play": Young Girls' Insider-Other Narratives of Sport', *Sociology of Sport Journal*, 22, pp. 158–77.

Copping, N. (2009) 'The Great Indoors: A Girls' Guide to Walking in High Heels', *FT Weekend Magazine*, August 8/9, p. 44.

Daly, A. (1987) 'Classical Ballet: A Discourse of Difference', *Women and Performance: A Journal of Feminist Theory*, 3 (2): 57–67.

Davis, K. (1997) 'Embody-ing Theory: Beyond Modernist and Postmodernist Readings of the Body', pp. 1–27 in *Embodied Practices: Feminist Perspectives on the Body*, K. Davis (ed.) (London: Sage).

Deem, R. (1986) *All Work and No Play? The Sociology of Women and Leisure* (Milton Keynes: Open University Press).

Downs, D. M., James, S. & Cowan, G. (2006) 'Body Objectification, Self-Esteem, and Relationship Satisfaction: A Comparison of Exotic Dancers and College Women', *Sex Roles*, 54, pp. 745–52.

Dryden, C. (1999) *Being Married, Doing Gender* (London: Routledge).

Duits, L.& van Zoonen, L. (2006) 'Headscarves and Porno-Chic: Disciplining Girls' Bodies in the European Multicultural Society', *European Journal of Women's Studies*, 13 (2), pp. 103–17.

Duits, L. & van Zoonen, L. (2007) 'Who's Afraid of Female Agency? A Rejoinder to Gill', *European Journal of Women's Studies*, 14 (2), pp. 161–70.

Smith, M. A. & Kollock, P. (eds) (1999) *Communities in Cyberspace* (London: Routledge).

Smyth, L. (2008) 'Gendered Spaces and Intimate Citizenship: The Case of Breastfeeding', *European Journal of Women's Studies*, 15, pp. 83–99.

Sparkes, A. C. (1995) 'Living Our Stories, Storying Our Lives, and the Spaces Inbetween: Life History Research as a Force for Change' in *Research in Physical Education and Sport: Exploring Alternative Visions*, A. C. Sparkes (ed.) (Lewes: Falmer Press).

Spender, D. (1995) *Nattering on the Net: Women, Power and Cyberspace* (Melbourne: Spinifex).

Stanley, L. & Wise, S. (1983) *Breaking Out Again: Feminist Ontology and Epistemology* (London: Routledge).

Steele, V. (1996) *Fetish: Fashion, Sex and Power* (New York: Oxford University Press).

Strunk, W. & White, E. B. (2000) *The Elements of Style*, Fourth edition (Needham Heights: Macmillan).

Tasker, Y. (1993) *Spectacular Bodies: Gender, Genre and the Action Cinema* (London: Routledge).

Tate, S. (1999) 'Making Your Body Your Signature: Weight Training and Transgressive Femininities' in *Practising Identities*, S. Roseneil & J. Seymour (eds) (London: Macmillan).

Tedlock, B. (2005) 'The Observation of Participation and the Emergence of Public Ethnography' in *The Sage Handbook of Qualitative Research*, N. K. Denzin & Y. S. Lincoln (eds) (London: Sage).

Thiel Stern, S. (2007) *Instant Identity: Adolescent Girls and the World of Instant Messaging* (New York: Peter Lang).

Thomsson, H. (1999) 'Yes, I Used to Exercise But ... – A Feminist Study of Exercise in the Life of Swedish Women', *Journal of Leisure Research*, 31 (1), pp. 35–56.

Truss, L. (2003) *Eats, Shoots & Leaves: The Zero Tolerance Approach to Punctuation* (London: Profile Books).

Tyler, I. (2000) 'Reframing Pregnant Embodiment' in *Transformations: Thinking Through Feminism*, S. Ahmed, J. Kilby, C. Lury, M. McNeil & B. Skeggs (eds.) (London and New York: Routledge).

Tyler, I. (2001) 'Skin-Tight: Celebrity, Pregnancy and Subjectivity' in *Transformations: Thinking Through the Skin*, S. Ahmed & J. Stacey (eds.) (London and New York: Routledge) pp. 69–83.

Ussher, J. M. (1997) *Fantasies of Femininity: Reframing the Boundaries of Sex* (London: Penguin Books).

Walkerdine, V. (2003) 'Reclassifying Upward Mobility: Femininity and the Neo-liberal Subject', *Gender and Education*, 15 (3), pp. 237–48.

Waskul, D. & Vannini, P. (eds) (2006) *Body/Embodiment: Symbolic Interaction and the Sociology of the Body* (Aldershot: Ashgate).

Waskul, D., Vannini, P. & Wisen, D. (2007) 'Women and their Clitoris: Personal Discovery, Signification, and Use', *Symbolic Interaction*, 30 (2), pp. 151–74.

Watson, B. & Scraton, S. (2001) 'Confronting Whiteness? Researching the Leisure Lives of South Asian Mothers', *Journal of Gender Studies* 10 (3), pp. 265–77.

Watson, J. (1998) 'Running around Like a Lunatic: Colin's Body and the Case of Male Embodiment' in *The Body in Everyday Life*, S. Nettleton & J. Watson (eds) (London: Routledge).

Wearing, B. (1995) 'Leisure and Resistance in an Ageing Society', *Leisure Studies*, 14, pp. 263–79.
Wedgwood, N. (2004) 'Kicking Like a Boy: Schoolgirl Australian Rules Football and Bi-Gendered Embodiment', *Sociology of Sport Journal*, 21, pp. 140–62.
Wesely, J. K. (2003) 'Exotic Dancing and the Negotiation of Identity: The Multiple Uses of Body Technologies', *Journal of Contemporary Ethnography*, 32 (6), December 2003, pp. 643–69.
Wheaton, B. (ed.) (2004) 'Introduction: Mapping the Lifestyle Sport-scape' in *Understanding Lifestyle Sports: Consumption, Identity and Difference* (London: Routledge).
Whelehan, I. (2000), *Overloaded: Popular Culture and the Future of Feminism* (London: The Women's Press).
Whitehead, S. & Biddle, S. (2008) 'Adolescent Girls' Perceptions of Physical Activity: A Focus Group Study', *European Physical Education Review*, 14, pp. 243–62.
Whitty, M. & Carr, A. (2006) *Cyberspace Romance: The Psychology of Online Relationships* (Basingstoke: Palgrave Macmillan).
Wilkins, A. C. (2004) '"So Full of Myself as a Chick": Goth Women, Sexual Independence, and Gender Egalitarianism', *Gender & Society*, 18, pp. 328–49.
Wilson, E. (1985) *Adorned in Dreams: Fashion and Modernity* (London: Virago).
Willson, J. (2008) *The Happy Stripper: Pleasures and Politics of the New Burlesque* (Oxford: I. B. Tauris).
Woods, P. (1999) *Successful Writing for Qualitative Researchers* (London: Routledge).
Wright, J. & Dewar, A. (1997) 'On Pleasure and Pain: Women Speak Out About Physical Activity' in *Researching Women and Sport*, G. Clarke & B. Humberstone (eds) (Basingstoke: Macmillan).
Young, I. M. (1990) *Throwing Like a Girl and Other Essays in Feminist Philosophy and Social Theory* (Bloomington: Indiana University Press).
Zaman, H. (1997) 'Islam, Well-Being and Physical Activity: Perceptions of Muslim Young Women' in *Researching Women and Sport*, G. Clarke & B. Humberstone (eds) (London: Macmillan).

Index